RICE TUNGRO DISEASE EPIDEMIC—AN ANALYSIS

RICE TUNGRO DISEASE EPIDEMIC—AN ANALYSIS

M. K. Satapathy
Department of Botany
Regional Institute of Education
Bhubaneswar 751022
and
A. Anjaneyulu
Former Head
Division of Plant Pathology
Central Rice Research Institute
Cuttack 753006

2001
DISCOVERY PUBLISHING HOUSE
New Delhi

First Published-2001

ISBN 81-7141-621-7

Published by :

DISCOVERY PUBLISHING HOUSE
4831/24, Ansari Road, Prahlad Street,
Daryaganj, New Delhi-110002 (INDIA)
☎ : 3279245 • Fax : 91-11-3253475
E-mail : dphtemp@indiatimes.com

Printed at :
ARORA OFFSET PRESS
DELHI-110092

Contents

Foreword

Advances in rice production and productivity are vital for national food security. Unlike wheat, rice suffers from serious pest and disease problems. Among the diseases Tungro is a very important one. This is why this book on Rice Tungro disease is a timely and valuable contribution.

We owe a deep debt of gratitude to Dr. M.K Satapathy and to the late Dr Anjaneyulu for their incisive analysis on an important disease.

This is the most comprehensive publication so far on Tungro disease. I am confident it will be of great value to scientists, scholars and extension workers.

M S Swaminathan

M S Swaminathan Research Foundation

Chennai

Preface

Tungro is one of the most important diseases of rice and a major constraint to stable rice production in South and South-east Asia, the rice bowl of the world. The disease causes significant yield losses sometimes reaching upto 100 per cent depending on the variety being cultivated and the time of infection. Besides economic importance, its aetiology and virus-vector relationships are unique which has attracted the attention of plant molecular biologists and biotechnologist today. The disease is caused by two distinct virus particles, one with RNA and the other with DNA genome, and both particles are transmitted in a semi-persistent manner by the green leafhopper. This RNA-DNA virus complex, their semi-persistent transmission and interaction in tungro disease syndrome are of rare examples in the field of plant virology.

Since the time of identification of tungro disease in 1960's, a lot of information has been gathered on various aspects of the disease and we have put them systematically in our monograph "Rice Tungro".

It is very surprising to note that the disease is still a mystery for plant pathologists, entomologists and rice scientists because of its unpredictable nature and uncertain outbreak. The disease appears suddenly, turns into an epidemic and vanishes thereafter. In the present book we have attempted to analyse various aspects of tungro epidemics and its unpredictable nature. We have discussed issues relating to Tungro epidemics based on the information available in the literature as well as from our experience on tungro research.

We have put the information in separate chapters such as factors affecting tungro epidemics, population dynamics of leaf-hopper vectors, tungro disease in different rice ecosystems, environment affecting tungro risk factors in tungro epidemics, diseases forecasting and finally its strategic management followed by references. In each chapter wherever it is felt, we have raised and discussed important issues to draw the attention of present and future researches. We sincerely hope that the information put in the present book shall be useful to agricultural scientists, plant pathologists, entomologist and extension workers interested in rice research besides students and teachers of plant pathology, general agriculture and botany.

We express our deep sense of gratitude to Dr. R.S. Paroda, DG, Indian Council of Agricultural Research : Dr R. Reddy, Head, Plant Pathology Division, Central Rice Research Institute, Cuttack; Dr. A.P.K. Reddy Head, Plant Pathology Division, Directorate of Rice Research, Hyderabad; Prof. D.K. Bhattacharjee, Prof. M.A. Khader, Prof. A.L.N. Sharma, and Prof. M.P. Sinha of RIE, Bhubaneswar for their inspiration and support for preparing this book.

We are thankful to Drs. H. Hibino and H. Koganezawa, National Agriculture Research Centre, Japan; Dr. Paul S. Teng, Dr. G.S Khush, International Rice Research institute, Philippines, Dr. J.M. Thresh, and Dr. T.C.B. Chancellor of National Resources Institute, University of Greenwich UK and Dr. R. Sridhar Central Rice Research Institute, Cuttack for sharing valuable discussions from time to time before and during preparing this book. We are extremely grateful to Dr. M.S. Swaminathan, Former Director-General, International Rice Research Institute, Philippines and presently UNESCO Professor of Ecotechnology and Chairman, M.S. Swaminathan Foundation, Chennai for kindly writing foreword for the book.

We express our sincere gratitude to editors of different scientific journal(s) for their kind permission to produce some of their published materials in this book.

M. K. Satapathy
A. Anjaneyulu

Introduction

Rice (*Oryza sativa L.*) is the most important food crop of the developing world. It is the primary staple for more than 2 billion people in Asia, the world's most densely populated region, and for hundreds of millions of people in Africa and Latin America. Worldwide rice is harvested on 148 million hectares representing more than 10 per cent of the earth's arable land. Total annual production is about 520 million tones of unmilled rough rice.

Asia is home to 59 per cent of the world's population. The farmers of Asia till 90 per cent of the world's harvested rice area and account for 92 per cent of global rice production. Presently, the rice consuming population is growing at 2 per cent a year globally. In humid and subhumid Asia rice is the primary staple food population is expected to increase by 58 per cent over the next 35 years. Recent projections indicate a world rice need of about 765 million tones in 2025 - 70 per cent more rice than is produced today. Consequently it is said that annual output must increase by over 5 million tones a year just to keep pace with population growth.

Of the several pest and diseases infecting rice tungro disease is economically the most important one. It is widely considered as a major constraint to rice production especially in intensively cultivated irrigated and favourable rice ecosystem in many countries of south and south-east Asia.

Tungro is a composite disease caused by two independent virus particles, rice tungro spherical virus (RTSV) and rice tungro bacilliform virus (RTBV) (Cabauatan and Hibino, 1988). Recent molecular studies have established that RTSV is a RNA virus and RTBV has a double stranded DNA genome (Jones *et al.*, 1991). Both the viruses are transmitted in a semi-persistent manner by green leafhopper, *Nephotettix virescens* and other specises of minor importance. The semi-persistent nature of the virus-vector relationship to tungro viruses is unique among leafhopper transmitted viruses. Bacilliform virus alone produces moderate type of symptoms whereas spherical form causes mild stunting without any foliar discolouration. Both viruses together cause serious symptoms.

Besides its molecular biology and virus-vector relationship, the high profile of tungro is due to its uncertain appearance. The disease is found sporadically but occasional outbreaks damage vast tracts of rice causing near to famine condition, declines thereafter without any trace of tungro in subsequent years. The disease causes havoc among rice cultivators and stands as a challenge before scientists, plant breeders, planners and politicians. Though frequency of tungro outbreaks has declined since 1960's major epidemics still occur and the disease is endemic in certain pockets in South and South-east Asian countries.

Looking to recent past, in 1990 the disease affected 90,000 ha of rice crop in coastal Andhra Pradesh, Orissa and West Bengal in India. Major tungro outbreak occurred during 1995 wet season causing extensive damage to crop in Central Java, Indonesia. In 1998 the disease is reported to have spread for the first time to the State of Pubjab (India) affecting 40,000 ha. An estimate of the global economic impact of the disease placed average annual losses due to tungro as high as US \$ 1.5×10^9.

In the last 35 years after the identification of tungro, a lot of information have been gathered on various aspects of the viruses and the disease and we have put them systematically in the monograph "Rice Tungro" (Anjaneyulu *et al.*, 1995) . In the present publication authors have tried to analyse various issues and factors that contribute to tungro epidemics in various parts of the world. Authors have attempted to probe into the mysteries behind its unpredictable nature and causes of uncertain outbreak. Besides disease spread authors have discussed risk factors associated with tungro disease, disease forecasting and its strategic management.

2

Economic and Academic Importance

Rice tungro disease (RTD) is the most economically important and wide spread disease of several virus diseases infecting rice known so far. This disease unknown in 1950's and considered as a nutritional disorder has become a major problem now. The introduction of short duration high yielding rice varieties such as TNI and IR8 which were susceptible to tungro viruses, and their wide and intensive cultivation in 1960's in Asian countries to maximise production has magnified the devastative nature of tungro. The disease is now widely distributed in South and South-east Asian countries, and its sudden and catastrophic epidemics in the past has led to great hardship and in some instances famine in the whole region. In the last three decades or more, RTD has been widely recognised as a serious constraint to rice production mostly in intensively cultivated irrigated areas in a number of countries in South and South-east Asia. (Hibino *et al.*, 1991; Chancellor and Thresh, 1997).

The potential of tungro to cause severe yield loss and lack of any effective control measure accounts for the high profile of tungro. The yield loss depends on the susceptibility of the cultivars as well as their growth stages at which virus infection occurs. (Jhon and Ghosh, 1980a). In early infection, yield loss in highly susceptible cultivars sometimes reaches upto 100 per cent (Rao and Anjaneyulu, 1980) and thus leads to complete crop failure. According to an estimate (Herdt, 1988) annual loss due to tungro alone stands above US $ 1.5 x 10^9 presently.

RTD is caused by a combination of two distinct viruses (Hibino *et al.*, 1978 a,b), rice tungro spherical virus (RTSV) and rice tungro bacilliform virus (RTBV). RTSV particles are isometric and 30 nm in diameter whereas RTBV particles are 100-300 nm in length and 30-35 nm in width (Plate-1). Both viruses multiply independently in the rice plant (Hibino, 1989). RTSV has a single stranded polyadenylated RNA of abut 12.2 kb as its genome (Hull *et al.*, 1991; Jones *et al.*, 1991; Shen *et al.*, 1993; Hull *et al.*, 1997). Its RNA sequence shows that there is a large open reading frame (ORF) capable of encoding a protein of 390 KDa and two short ORFs at the 3'end. RTBV has a circular double stranded DNA genome of 8.0 kbp with two discontinuities at specific sites, one in each strand (Hay *et al.*, 1991; Qu *et al.*, 1991; Kano *et al.*, 1992). Sequencing of RTBV DNA suggests four ORFs encoding proteins of 24 KDa, 12 KDa and 46 KDa.

Rice tungro viruses are transmitted by green leafhopper, *Nephotettix virescens* (Distant), *N. nigropictus* and few others of insignificant importance (Anjaneyulu *et al.*, 1994). Most of the leafhopper transmitted viruses are propagative or circulative in their insect vectors whereas in tungro the virus-vector relationship is non-persistent (Ling, 1966) an unique feature among leafhopper transmitted viruses with the exception of maize chlorotic dwarf virus. RTSV is transmitted independently whereas transmission of RTBV is dependant on RTSV. The association of two different virus particles (isometric and bacilliform) in the tungro disease syndrome and their relationship with the insect vector are of great academic interest. These peculiar characteristics such as particle type, genome type, host-virus and virus-vector relationship signify the importance of tungro in the field of plant virology, molecular biology and entomology besides its economic importance.

3

Tungro Epidemics: Facts and Figures

Rice tungro disease is widely distributed in almost all South and South-east Asian countries where rice is predominately cultivated. The disease is known by different local names in different countries.

(A) INDIA

Soon after detection of tungro disease from the experimental farm of International Rice Research Institute, (IRRI), Philippines, tungro was reported as "leaf yellowing" transmitted by *N. virescens* from the state of West Bengal in India (Raychaudhuri *et al.,* 1967). After John (1968) described it as tungro, a major epidemic was recorded in the wet season of 1969 from the states of Bihar, Uttar Pradesh and West Bengal (John, 1970) where two high yielding varieties, 'Padma and Jaya' were cultivated. Tungro epidemics then appeared sporadically in Kerala during 1973 and Assam and Tripura during 1974 (Anjaneyulu and Chakrabarti, 1977). Thereafter the prevalence of the disease has increased in the eastern and Southern states where rice is grown extensively throughout the year. A severe epidemic was observed during 1981 kharif season in the states of Bihar and West Bengal affecting 40-100 per cent of the crop (Singh *et al.,* 1982). During 1984-

85 the disease attacked 80,000 ha of rice planted mostly with high yielding varieties in Andhra Pradesh and Tamil Nadu (Chowdhury, 1997). In 1990 about 90,000 ha of rice in the states of Andhra Pradesh, Orissa, West Bengal and Assam were affected by tungro (Saikia *et al.*, 1992; Anjaneyulu *et al.*, 1994). In 1994 the disease again appeared in West Bengal. During 1996-98, a disease showing tungro like symptoms called as yellow stunt syndrome appeared in Punjab affecting about 40,000 ha of rice fields (Azzam *et al.*, 1999).

Other South and South-East Asian Countries

Besides India, the disease has been recorded from many South and South-east Asian countries such as the Philippines, Indonesia, Malaysis, Thailand, Bangaladesh, China, Pakistan, Nepal, Sri Lanka, Vietnam and Japan of which RTD often appears in epidemic form in the first five countries. In Japan a mild form of tungro disease known as Waika (Saito, 1977) has been observed.

Though tungro was reported from IRRI farm during 1963, this disease is said to have occurred in 1940s in the Philippines. Tungro is endemic to the Philippines and major epidemics have appeared in 1957, 1962, 1969, 1971, 1975, 1977 and 1983-84. Serrano (1957) observes that '*accep na pula*' or stunt disease which is similar to tungro was extremely destructive in 1940's thoughout the major rice growing regions of the Philippines and caused 30 per cent loss overall, equivalent to 1.4 million t. each year.

In 1970 a sudden epidemic of tungro broke out (Ou *et al.*, 1974) in provinces of central Luzon and Cotabato affecting about 100,000 ha of rice field planted with high yielding varieties. The disease continued for next two years affecting 70,000 ha in 1971 and 40,000 ha in 1972. This brought drastic reduction in rice yield with a loss of about 31.15 million pesos. During 1984, major rice growing regions including the central plain of Luzon and parts of Visayas and Mindanao islands were severely affected by tungro when IR8, IR36 and IR42 were planted widely in the Philippines. In 1987 tungro affected 5 of 12 regions of the country affecting 11, 164 ha (Reyes, 1989). In 1989 dry season large areas were affected by tungro in the Bicol region where major varieties grown by farmers were IR36, IR42, IR60, IR64, IR72 and BPI Ri10. Between 1988 and 1993, 2200 and 2100 ha were affected by tungro in North Cotabato and Davao provinces in the Mindanao region respectively. In Davao del Norte during 1993, tungro incidence was the highest affecting 50328 cavans of rice amounting to 10588.853 pesos. In the same year RTD incidence ranged from 34 to 90 per cent in the field in the Davao del sur region (Baria, 1997) where varieties such as IR60, IR64, IR72 were mostly planted.

Penyakit merah known in Malaysia since 1934 (Singh, 1967) was confirmed to be tungro in the year 1965 (Ou, 1965; Ou and Goh, 1966). In 1964 first tungro epidemic appeared in Krian district of Perak state affecting 10,000 ha and subsequently in 1969 (Lim, 1972). However after 1980, tungro was reported from several localities in the states of Penang, Kedah and Perlis and in 1982 more than 17,507 ha of paddy were affected by the disease. The yield loss in 1982 was estimated to be US $ 8.8 million. In 1981 tungro occurred mainly in the Southern part of MADA scheme where 5884 ha was affected. In 1982 the disease severely damaged 5839 ha and reached a maximum of 8655 ha in 1983. He value of crop less for 1981-83 was estimated to be US $ 10 million. (Chen and Jatil, 1997). following nation wide campaign to control tungro disease, the area affected by tungro in Peninsular Malaysia was reduced to 12,500 ha in 1983, 3100 ha in 1984 and 1000 ha in 1985. From 1986 to July 1989 the area affected by tungro was about 330 ha. In 1990 there was a resurgence of tungo disease affecting. 1879 ha (Chen and Othman, 1991). Following use of control measures such as insecticide application, roguing, destruction of stubbles etc. the affected area was greatly reduced to only 67 ha in 1992, 24 ha in 1993 and 20 ha in 1994 (Chen and Jatil, 1997).

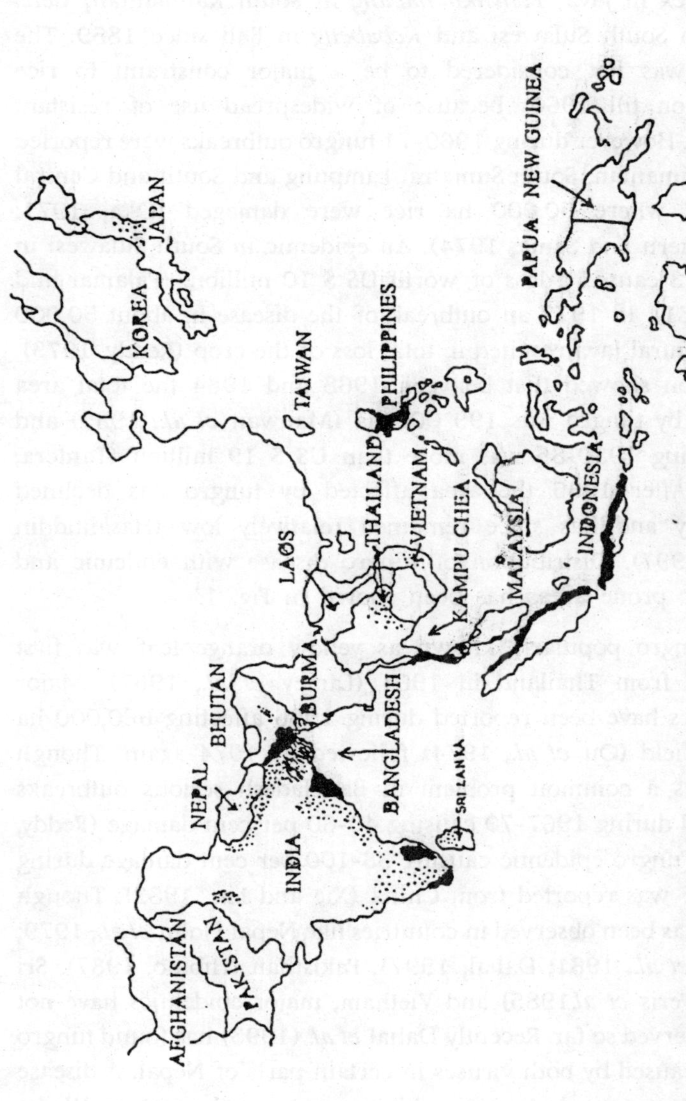

Fig.1. Distribution of rice tungro dieseae with endemic (dark shaded) and epidemic (light shadow) areas in South and South East Asia.

Tungro is known by different local names in Indonesia such as *Mentek* in Java, *Penyakit habang* in South Kalimantan, *Cella pance* in South Sulawesi and *Kebabeng* in Bali since 1859. The disease was not considered to be a major constraint to rice production till 1960s because of widespread use of resistant varieties. However during 1969-71 tungro outbreaks were reported from Kaimantan, South Sumatra, Lampung and South and Central Sulawesi where 50,000 ha rice were damaged (Oka, 1971; Van Haltern and Sama, 1974). An epidemic in South Sulawesi in 1972 -73 caused a loss of worth US $ 10 million. (Palamar and Rao, 1981). In 1972 an outbreak of the disease in about 50,000 ha in Central Java resulted in total loss of the crop (Reddy, 1973). Estimation showed that between 1968 and 1984 the total area affected by tungro was 199 000 ha (Manwan *et al.*, 1985) and loss during 1980-85 was more than US $ 19 million (Tantera, 1985). After 1985 the area affected by tungro has declined gradually and has since remained relatively low (Hasanuddin *et al.*, 1997). Distribution of tungro disease with endemic and epidemic prone areas has been shown in Fig. 1.

Tungro popularly known as yellow orange leaf was first reported from Thailand in 1964 (Lamey *et al.*, 1967). Major epidemics have been reported during 1966 affecting 660,000 ha of rice field (Ou *et al.*, 1974) followed by 1974 again. Though tungro is a common problem in Bangladesh serious outbreaks occurred during 1967-70 causing 40-60 per cent damage (Reddy, 1973). Tungro epidemic causing 38-100 per cent damage during 1980-81 was reported from China (Xie and Lin, 1982). Though tungro has been observed in countries like Nepal (John *et al.*, 1979; Omura *et al.*, 1981; Dahal, 1997), Pakisthan (Hibino, 1987), Sri Lanka (Peris *et al.*1985) and Vietham, major epidemics have not been observed so far. Recently Dahal *et al.* (1993) has found tungro disease caused by both viruses in certain parts of Nepal. A disease similar to tungro but very mild in symptoms known as Waika occurs in Japan. (Saito, 1977). A major epidemic of Waika occurred in Kyushu area of Japan during 1973 affecting 24,825 ha and causing a loss of 10,000 tons in grain yield (Shinkai, 1977 a,b). The damage due to tungro epidemics in different countries is shown in table 1.

Table 1 : Outbreaks of tungro and tungro like disease epidemics in different countries with damage

Country	*Year*	*Region/State*	*Damage*
India	1969	Bihar, Uttar Pradesh, West Bengal	40-50%
	1973	Kerala	
	1974	Assam and Tripura	
	1981	Bihar and West Bengal	40-100%
	1984	Andhra Pradesh	
	1985	Andhra Pradesh and Tamil Nadu	80,000 ha
	1990	Andhra Pradesh, Orissa, West Bengal & Assam	90,000 ha
	1999	West Bengal	
	1997-98	Punjab	40,000 ha
Indonesia	1969-71	Kalimantan, South Sumatra, Lampung & South Sulawesi	50,000 ha
	1972	Central Java	50,000 ha
	1972-73	South Sulawesi	97.254 ha
	1980-83	Bali	16,000 ha
	1981-83	Java	12,316 ha
	1982-83	Kalimantan & Sumatra	8,015 ha 4109 ha
	1984-92	Sulawesi, Bali & Java	18,985 ha 16481 ha 11837 ha

(contd...)

1	*2*	*3*	*4*
Philippines	1970	Central Luzon & Cotabato	100,000 ha
	1971	do	70,000 ha
	1972	do	40,000 ha
	1984	Central Luzon, Visayas & Mindanao	80,000 ha
	1987	Bivol, Vrnytsl Luzon, Visayas & Cotabato	11,164 ha
	1989	Bicol	
	1988	North cotabato	2200 ha
	1993	Davao	2100 ha
	1993	Davao del sur	34-90%
Malaysia	1964	Perak state	10,000 ha
	1982	Penang, Kedah, Perlis	17507 ha
	1981	Southern part of MADA scheme	5884 ha
	1982	-do-	5839 ha
	1983	-do-	3100 ha
	1984	Entire country	1000 ha
	1990	do	1879 ha
Thailand	1965	Central plane	50,000 ha
	1966	Central plane	660,000 ha
Bangaladesh	1967-70		40-60%
China	1980-81		38-100%
Nepal	1993	Harinath	28-30%
		Pawanipur	70-80%

Symptoms, Transmission and Disease Identification

Tungro is the most economically important rice virus disease in South and Southeast Asia. It is composite disease caused by two distinct viruses, rice tungro bacilliform (RTBV) and rice tungro spherical (RTSV) viruses. (Hibino *et al.*, 1978a, 1979).

Tungro disease symptoms are often confused with insect infestation and nutritional disorders such as nitrogen and iron deficiency. Tungro disease causes discolouration of leaves, stunted growth, reduction in tiller numbers and delayed flowering (Ling, 1972, Anjanejulu *et al.*, 1994) (Plate-2). Tungro diseased leaves look yellow/orange in colour, sightly rolled outward and somewhat spirally twisted. The discolouration starts form the tip of the leaf and extends to the lower part of the leaf blade. Young leaves may have a mottled appearance and old leaves show rusty specks of various sizes (IRRI, 1992). In susceptible cultivars, symptoms persist throughout the plant growth whereas moderately resistant cultivars show partial discolouration and slowly recover from symptoms and appear green in later stages of crop growth. The disease induces delayed flowering. Panicles are often small and sterile. Early infection in susceptible cultivars sometimes brings premature death to rice plants (Ou, 1985). Symptoms of tungro disease are clearly visible in plate 1.

The susceptible plants do not flower after infection whereas resistant cultivars exhibit delayed flowering. The percentage of infection decreases with increase in plant age at the time of infection. (Ling and Palomar, 1966). Early infection reduces the rate of tillering but it (tillering) remains the same when infection occurs at an advanced stage. With increase in plant age the latent period of virus infection increases (Astika *et al.*, 1992).

Disease symptom is influenced by rice cultivars, their growth stages, virus particle(s) present, strain(s) of the virus and environmental condition such as host nutrition. (Satapathy *et al.*, 1998). The symptoms become one pronounced with nitrogen deficiency. With increase in nitrogen fertilizer, the symptoms get masked temporarily. In mature plants the discolouration does not become pronounced. Insect vectors such as *Recilia dorsalis* (Motsch.) and *N. nigropictus* (Stal.) while transmitting the tungro viruses procedure very mild symptoms (Hino *et al.*, 1974) due to unknown reasons. RTBV alone causes yellowing and stunting symptoms that are enhanced when RTSV accompanies RTBV. RTSV alone does not induce distinct symptoms except mild stunting (Hibino *et al.*, 1978a). RTSV alone also does not cause much yield losses. Disease severity besides plant age depends on the total amount of virus concentration in the plant tissues. Though RTSV generally does not induce any clear symptoms in plant tissues, in *Oryza glaberrima*, it causes stunting, development of pale green and narrow leaves and short panicles with few grains which are rare characteristics (Cabauatan and Koganezawa 1992).

Sometimes vectors alone (non-viruliferous green leafhoppers) feeding on rice plants cause yellowish/brownish discolouration of leaves, reduction in plant height/tiller number due to sucking of plant sap and subsequent plugging of conducting vessels. However the plants mostly recover, when the insect population comes alone.

Transmission

Tungro disease is not transmitted by seeds, mechanical contact or soil. The viruses are exclusively transmitted by rice green leafhoppers, *Nephotettix virescens, N. nigropictus, N. Malayanus, N. parvus* and *Recilia dorsalis* of which *N. virescens* is the most efficient vector (Ling, 1970; Rivera and Ou, 1965) followed by *N. nigropictus.* Though both adults and nymphs transmit viruses, females are more efficient than males and nymphs are less efficient than adults in virus transmission (Anjaneyulu, 1975a; ling, 1975a). Nymphs of all growth stages transmit the viruses, the third, fourth and fifth instar nymphs being more efficient then first and second instars. However from Thailand Hino *et al.* (1974) had noted nymphs to be more efficient than adults.

RTSV can be transmitted independently by the insect vector but transmission of RTBV either alone or with RTSV depends on the presence of RTSV. (Hibino *et al.*, 1978a, b; Cabauatan and Hibino, 1985). RTBV transmission by the leafhopper occurs only when the leafhopper is previously or simultaneously fed on RTSV diseased plants. The efficient vector *Nephotettix virescence* can transmit both (RTBV+RTSV) or single (RTBV/RTSV) particle alone from diseased plants harbouring both the viruses to health rice plant. Nymphs behave in the same way like adults.

There is no incubation period in the vector and the insects transit the viruses immediately after acquisition feeding. The minimum acquisition feeding period for RTSV and RTBV are 10 and 30 minutes respectively for *N. virescens* (Hibino *et al.*, 1979).The minimum feeding period for virus transmission is 30 seconds (Dahal *et al.*, 1990b). *N. virescens* (plate-3) is both a xylem and phloem feeder on rice. On leafhopper susceptible cultivars, it feeds from the phloem and transmits both the viruses (RTSV+RTBV), whereas in leafhopper resistant cultivars, the feeding is mainly from the xylem and during feeding it transmits predominantly RTBV alone (Dahal *et al.*, 1990b; Heinrichs and

Rapusas, 194). Compared with *N. virescens, N. nigropictus* transmits both the viruses less efficiently during dual transmission but RTSV efficiently from RTSV source. *N. nigropitus* prefers to feed on weeds and consequently transmits the viruses from rice to weeds. The other three (*N. malayanus, N. parvus* and *R. dorsalis*) species are poor transmitters of tungro viruses and their transmitting efficiency is less than 10 per cent. Consequently their potentiality as vectors of viruses is not very clear.

Disease Identification

Since tungro disease symptoms are often confused with many disorders induced by biotic and abiotic factors, various diagnostic techniques such as symptomatology, chemical tests, transmission tests, serology (Latex–Agglutination test, Enzyme linked immunosorbent assay), polymerase chain reaction, parafilm mini-ELISA, rapid immuno filter paper assay, enzyme dot blot assay, use of indicator plant, electron microscopy etc. have been used for disease identification (Arti *et al.*, 1997: IRRI, 1983; Cabunagan *et al.*, 1987; Cabauatan *et al.*, 1993; and Cabauatan Koganezawa, 1997). Each technique has some advantages as well as limitations, Various techniques of identification have been described in our earlier publication (Anjaneyulu *et al.*, 1994)

In developing countries, symptomatology and transmission tests are commonly used for disease diagnosis. Generally tungro disease induces discolouration (yellow or orange colour) of leaves, stunted growth and reduction in tiller numbers. If a plant is suspected to be tungro infected, it may be uprooted and transferred into a pot with soil and fertilizer. After few days, if the plant recovers from the symptoms, it is not tungro. However if the symptoms appear in young leaves, it is tungro infected.

In transmission test, non-viruliferous leafhoppers are forced to feed on diseased plants and are then transferred to healthy seedlings kept in test tubes for two to three days for inoculation feeding. Inoculated seedlings collected and then transplanted, if show symptoms after about a week, the original plant is said to be tungro diseased.

Iodine test is also a commonly used practice for tungro disease identification. In this test the second leaf of a tiller from a suspected plant is cut and put in iodine solution. Black colouration at the cut edge of the leaf shows the leaf to be infected. However, brown colouration at the cut end shows the leaf to be healthy. Scoring for the reaction is done 15-20 min after dipping the leaf in iodine solution.

Latex agglutination test is a simple but reliable serological test often employed for tungro identification. In this method equal volume of plant sap (collected after grinding with 0.05m Tris buffer and centrifugation) mixed with antiserum-treated latex suspension. The mixture is then shaked thoroughly for 30-40 minutes. Examination of one/two drops of the mixture under microscope if shows signs of clumping, presence of virus (consequently infection) is confirmed.

Enzyme linked immunosorbent assay (ELISA) has been the most commonly used practice for routine detection of tungro viruses of many samples at a time. A test sample containing specific antigen (tungro viruses) is incubated with the antibody sensitized solid phase. After few hours, of entrapment of the antigens by the antibodies enzyme-labelled antibodies are added after removal of excess or unfixed antigens. The labeled antibodies bind to the antigens already fixed to the coating antibodies forming a double sandwich. The enzyme substrate is added and after incubation, the reaction is stopped and read in a spectrophotometer at 405 nm in an ELISA reader. The colour change in the substrate is proportional to the amount of enzyme present that is again proportional to the concentration of the antigen bound (Baiet *et al.*, 1985). Though infected plants showing no foliar symptoms can be identified by this technique, it is an expensive method and requires laboratory equipment that is usually not readily available in laboratories and also it cannot be carried out in the field.

Consequently, a modification of conventional ELISA known as parafilm mini-ELISA (PAM-ELISA) has been developed. It involves (Cabauatan and Koganezawa, 1997) the use of a parafilm membrane as the solid phase instead of microlitre plates. A small piece of parafilm is wrapped around a glass microscope slide. Small imprints that serve as mini wells are made by pressing one end of a glass rod. (2 mm d.) on to the membrane. The procedure follows the principle of double antibody Sandwich-ELISA. After coating with specific antibody, the sample and conjugate (alkaline phosphatase labelled specific antibody) are applied together. Nitroblue tetrazolium and 5-bromo-4-chloro-3-indolylphosphate are used as substrates. A blue colour reaction indicating a positive result develops within 10 minutes. All procedures are done at room temperature and membranes are incubated for 3h inside a covered plastic box lined with moist paper towel to prevent drying. Washing of membranes after each step is done by gently flooding the membrane with wash buffer using either a pasteur pipette or a wash bottle. The technique can detect tungro viruses in a 100 fold dilution of crude sap. The entire assay time is reduced to 4th and it does not require sophisticated equipment and can be readily adopted.

Besides PAM-ELISA, rapid immuno filter paper assay (RIPA) is also a sensitive method to detect rice tungro viruses. This method involves the use of two kinds of latex beads, white latex as the solid phase and pink latex as the tracer. The antibody coated white latex is immobilized on whatman glass filter paper strips (Whatman GF/A:0.5 x 9 cm.). A fine pointed brush is used to apply a thin coat of white latex suspension on the filter paper strips of 1.5 cm from the lower end. Latex coated filter paper strips are air dried and stored in a desiccator at room temperature. For virus assay 0.1 g of infected rice leaf is homogenized in 900 L of extraction buffer (TBS with 0.01 Na_2So_3) using a mortar and pestle. The extracts are clarified by centrifugation at 15,000 rpm for 10 min, 100 vl of clarified extract are placed in flat bottomed Eppendorf tubes in which the lower end of the latex coated filter paper strip is dipped until the extract is fully absorbed. Then 100 nl of IgG-coated pink latex diluted to 0.025% (v/v) with TBS is added to the same tube. The pink latex suspension moves upward by capillary action and a pink band appears at the spot where the white latex was applied to indicate a positive reaction. The sensitivity of RIPA is comparable to ELISA for both RTBV and RTSV but it is simpler, less time consuming and cheaper than ELISA. Moreover RTBV and RTSV can be detected simultaneously in a single filter paper strip sensitized with antibody to both the viruses.

Nature of Tungro Epidemics Caused by Lone and Individual Factors

Rice Tungro Spherical virus

Two virus particles (RTSV and RTBV) causing tungro disease besides multiplying independently are transmitted in a semi-persistent manner by rice green leafhoppers. RTSV is transmitted independently but RTBV is transmitted only after the leafhopper acquires the virus from source plants infected either with both RTBV and RTSV or when they acquire RTSV first followed by RTBV (Cabauatan and Hibino, 1988). RTSV is transmitted more efficiently and cause latent infection.

RTSV is said to spread as an independent disease in the Philippines (Bajet *et al.*, 1986), China, Nepal (Dahal *et al.*, 1997) and probably in other Asian countries (Hibino, 1989). In the field RTSV infection proceeds RTSV+RTBV infection. Moreover single infections with RTSV alone reaches a peak nearly within a month after transplanting and declines thereafter as subsequent infection of RTBV becomes apparent and dual infection with RTSV and RTBV takes up (Tiongco *et al.*, 1993). As RTSV infected rice plants do not show any distinct symptoms, RTSV spread remains unnoticed in the field. However the steady increase of RTSV infection in rice fields in the early phase facilitates the subsequent acquisition and spread of RTBV by vector leafhopper and the resultant increase in tungro disease (Chancellor *et al.*, 1996b).

Interestingly it has been noted that *Oryza glaberrima* produces distinct symptoms after getting infected by RTSV. RTSV infected plants besides showing severe stunting, produce pale green, erect and narrow leaves and short panicles with grains of small size (Cabauatan and Koganezawa, 1993). RTSV has two strains. TKM-6 and some of its derivatives like IR20, IR26, IR30 etc. are resistant to typical and common strain of RTSV while TKM-6 is susceptible to atypical strain (Cabauatan *et al.*, 1995). Although RTSV causes mild symptoms, greenhouse experiments have shown that RTSV alone causes as high as 40 per cent yield loss in IR36, about 20 per cent in IR54, TN1 and less than 5 per cent in Balimau Putih, Sigadis and Utri Rajapan (Hasanuddin and Hibino, 1989). It may be interesting to look into the epidemiological studies of RTSV alone in high yielding semidwarf varieties, as many Gam Pai-30-12-15 progenies such as IR28, IR29, IR50, IR54, IR60 etc. show RTBV infection commonly but get infected with RTSV at relatively high rate when inoculated by leafhoppers that had fed on RTSV infected plants. Further under high insect and disease pressure the plants get infection from both RTSV and RTBV (Dahal and Hibino, 1985; Hibino *et al.*, 1988).

Rice Tungro Bacilliform virus

Most RTSV resistant varieties get infection from RTBV. TKM6 progenies such as IR20, IR26 and IR30 (Hibino *et al.*, 1990) and Gam Pai 30-12-15 progenies like IR28, IR29, IR34, IR50, IR54, IR60 etc (Daguioag *et al.*, 1984; Dahal *et al.*, 1988) show very low RTBV infection both in artificial and natural conditions. As RTBV alone does not spread secondarily in the field, primary infection remains almost the same and it is very effective in reducing disease spread in large size plots.

These RTSV resistant cultivars can provide an effective control of tungro in large fields as plants infected with RTBV do not serve as sources for secondary spread in the field. It is likely to have little effect in small plots as there is a probability of high influx of leafhopper vectors from neighbouring fields carrying both the viruses into the small plot. However inspite of the potentiality, these resistant sources may not be stable either due to arrival of a new strain of the virus through migratory leafhoppers or due to development of a new strain of the virus due to wide cultivation of a variety.

Ratna, a derivative of TMK6 is widely cultivated in eastern India. It does not show clear-cut symptoms with tungro and virus has not been recovered by back indexing through non-viruliferous leafhoppers. The limited stunting and disease spread in Ratna and Annapurna are attributed to the presence of RTBV.

The vector, Nephotettix spp

Rice tungro is transmitted by five green leafhopper species, *Nephotettix virescens*, *N. nigropictus*, *N. cincticeps*, *N. malayanus* and *N. parvus* of which the first three species are very important.

Nephotettix virescens

N. virescens is monophagous and it's most preferred host is rice (Viswanathan and Kalode, 1981; Siwi *et al.*, 1987). It is widely distributed in rice fields in Burma, China, Malaysia, Indonesia, Vietnam, Thailand, the Philippines, Pakistan, Sri Lanka, Hong Kong, Nepal and Southern parts of Japan. Because of its early colonization of rice crops and, more rapid population development and higher transmission efficiency. *N. virescens* is considered to be the most efficient vector of tungro viruses (Shukla and Anianeyulu 1982c; Siwi *et al.*, 1987; Chancellor *et al.*, 1996b) and responsible for major outbreaks of tungro epidemics in Asian countries. Though many cultivars resistant to *N. virescens* have been developed to curtail the disease outbreak, most of them have suffered from the disease after a few years of intensive cultivation (Hibino and Anianeyulu, 1991) because of development of insect biotypes with higher virulence (Dahal *et al.*, 1990a). Further even resistant varieties suffer when vector pressure is very high in the field. Once the insect vectors shift their virulence to a resistant cultivar, all cultivars having the same resistance gene suffer from the disease.

N. nigropictus

N. nigropictus is seen in all the Asian countries along with *N. virescens* and also in Australia. *N. nigropictus* is differentiated from *N. virescens* by its dark pigmentation and the number of spines on the male aedeagus. *N. nigropictus* which is less efficient than *N. virescens* in virus transmission prefers weeds to rice and plays limited role in tungro epidemics. In single crop areas, it is reported to be responsible for carrying over of viruses from weeds to rice (Khan *et al.*, 1991) and helping in the initiation of the disease. Once the disease is initiated, further spread from plant to plant within a rice field occurs through *N. virescens*, the prime vector of rice tungro. In Kuttanadu areas of Kerala (a tropical Indian state), this species (*N. nigropitus*) was observed to be solely responsible for outbreak of tungro during 1974 (Anjaneyulu and Chakrabarti, 1977). Of course *N. nigropictus* being a poor transmitter, the fields were not completely infected.

N. cincticeps

N. cincticeps is distributed in Japan, China, Taiwan and certain parts of Soviet far east. In China and Nepal, tungro is transmitted by *N. cincticeps*. A mild form of tungro known as waika is also transmitted by it in Japan.

Thus the type and nature of tungro epidemic varies depending on dominant vector species in the location. the other two vector species, *N. malayanus* and *N. parvus* are insignificant in causing tungro epidemics as their population size is very low in rice fields besides transmission rate being very poor.

Simulation of Tungro Epiphytotics for Experimental Purpose

In non-endemic areas, tungro epidemics break out once in about 5 to 7 years interval and it stands as a limitation for experiemental studies. For conducting regular epidemiological experiments the authors have simulated epidemics in the experimental farm of Central Rice Research Institute (CRRI), Cuttack, Orissa, India taking into consideration important factors such as annual vector population build up, host susceptibility, early growth stages of rice crop and virus inoculum (Anjaneyulu 1975b).

For availability of vector *N. virescens*, responsible for disease spread, sowing and planting dates are delayed by four to six weeks to the normal planting time during kharif season. Sowing and planting of test materials are carried out in the middle of August and second week of September respectively for creating a good disease pressure at Cuttack. For preparing virus inoculum, 20-day old seedlings of susceptible cultivar Jaya, inoculated in the greenhouse are transplanted temporarily in the field for use. After two weeks infected tillers of Jaya showing clear symptoms are planted at regular intervals between test materials/experimental plots to initiate disease spread (Fig.2). Since susceptible cultivars have a profound influence on the disease spread, two lines of tungro susceptible TN1 alternating with two lines of test

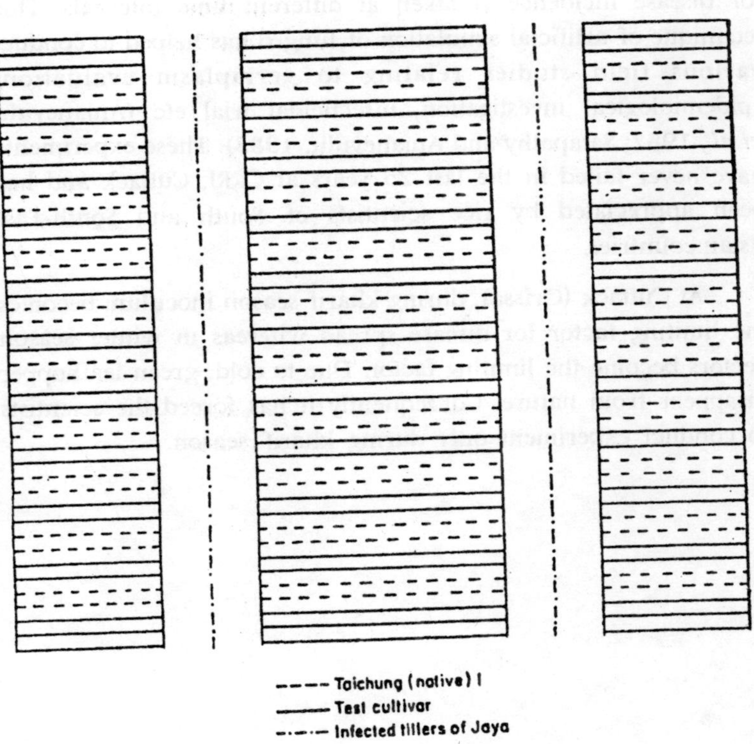

Fig.2. Layout of field screening method (Anjaneyulu *et al.* 1982a)

materials are planted for creating a disease pressure of epidemic proportion.

Through this method, 30 to 40 days after transplanting (DAT), 100 per cent infection occurs in susceptible varieties. Observation for disease incidence is taken at different time intervals. This technique of artificial simulation of tungro has helped to conduct various field studies relating to germplasm evaluation, epidemiological investigation, insecticidal trial etc (Anjaneyulu *et al.*, 1982; Satapathy and Anjaneyulu, 1983). These experiments have never failed in the last 25 years at CRRI, Cuttack and has been appreciated by rice scientists of South and South-East Asian countries.

At Cuttack (Orissa), during kharif season inoculum becomes the limiting factor for disease spread whereas in winter season vectors become the limiting factor. Due to cold. green leafhopper disappear from nature. Consequently it has forced the scientists to conduct experiment only during kharif season.

Spatio-Temporal Dynamics of Tungro Disease

Tungro is a polycyclic disease (Thesh, 1983; Anjaneyulu *et al.*, 1994). The disease is initiated either due to colonization of leafhopper vectors on the existing inoculum sources in the field carried over from nursery beds or because of immigration of viruliferous hoppers from outside sources into the healthy plots or both. Most of the disease spread is due to repeated plant to plant secondary spread in the field from available inoculum foci leading to multiplication of the pathogen through successive generation in the course of the epidemic. Consequently repeated infection cycles are observed in the life span of a rice crop. Local spread occurs from plant to plant by nymphs as well as adult leafhoppers whereas long distance spread from field to field takes place by migration of leafhoppers carrying inoculum.

Tungro disease spread over time shows a sigmoid curve (Fig.3). Disease progress picks up and becomes fast few days after initiation but gradually slows down as plants get matured and few healthy plants are left for further infection. The disease progress is described by logistic model (Madden *et al.*, 1990; Chancellor *et al.*, 1996a). Disease progress curves for tungro in certain rice varieties with different degrees of resistance has been shown in Fig.3.

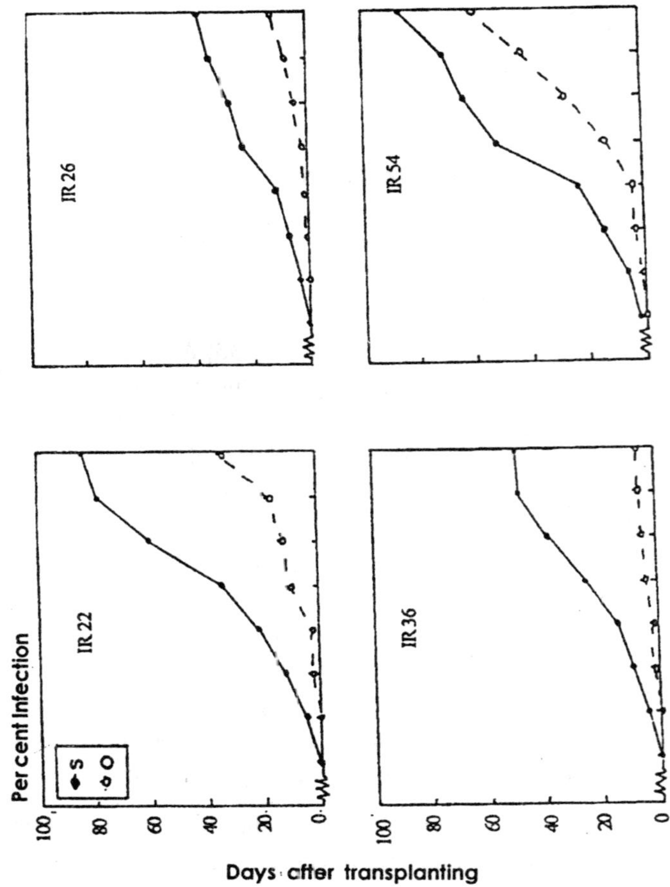

Fig.3. Progress curves of rice tungro disease in different varieties with (S) and without (O) a source of inoculum in experimental fields (Satapathy *et al.*, 1996 with modification)

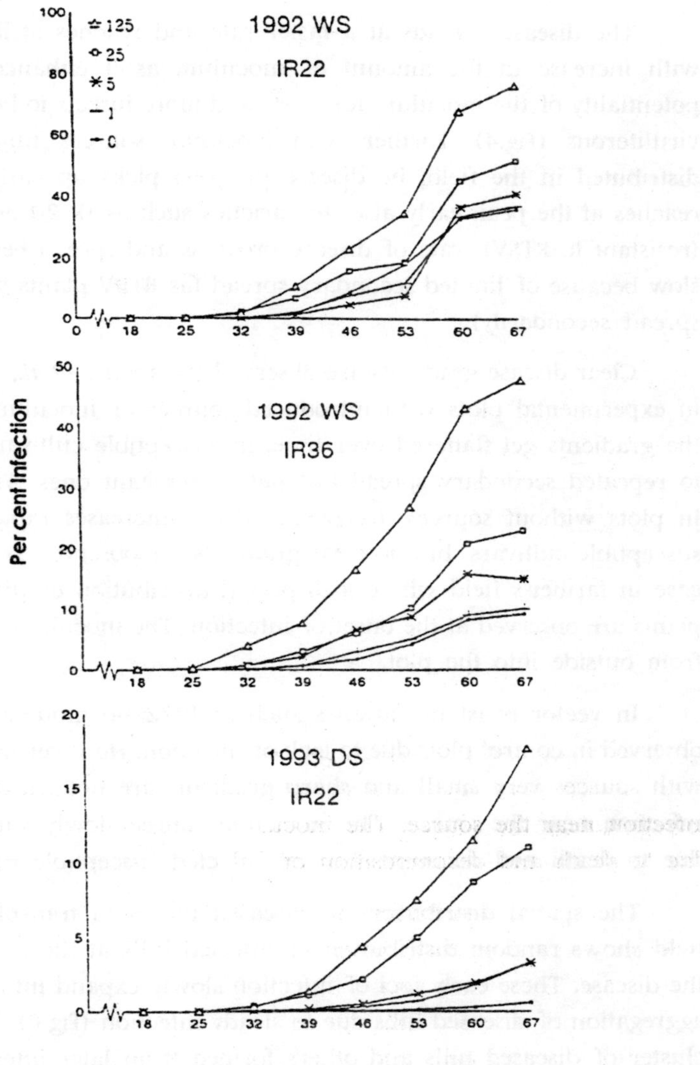

Fig.4. Disease progress curves of rice tungro disease in 1992 wet season and 1993 dry season in two different varieties with different amount of inoculum sources

The disease spreads at a faster rate and reaches at a peak with increase in the amount of inoculum as it enhances the potentiality of the inoculum for more and more insects to become viruliferous (Fig.4). Further with inoculum sources randomly distributed in the field, he disease progress picks up early and reaches at the peak early also. In varieties such as IR 20 or IR26 (resistant to RTSV), rate of disease progress and spread becomes slow because of limited secondary spread (as RTBV plants cannot spread secondarily).

Clear disease gradients are observed (Satapathy *et al.*, 1996) in experimental plots with introduced sources of inoculum and the gradients get flattered over time, in susceptible cultivars due to repeated secondary spread but not in resistant ones (Fig. 5). In plots without sources, tungro incidence increases rapidly in susceptible cultivars but no clear gradients are observed as is the case in farmer's field where a dispersed distribution of diseased plants are observed at the onset of infection. The inoculum comes from outside into the plot.

In vector resistant varieties such as IR72 no gradients are observed in control plots due to lack of infection. However in plots with sources very small and sharp gradients are noticed due to infection near the source. The inoculum source slowly vanishes due to death and decomposition of infected susceptible plants.

The spatial distribution of diseased plants in transplanted field shows random distribution of infected hills at the onset of the disease. These early foci of infection slowly expand into clear aggregation of diseased hills due to steady infection (Fig.6). These cluster of diseased hills and others formed from later infections at different points in the plot, continue to grow until they coalesce in the later stages reaching an uniform distribution. Spatio-temporal development of tungro in experimental plots of IR22, IR26, IR36 and IR72 has been shown in Fig. 7-10.

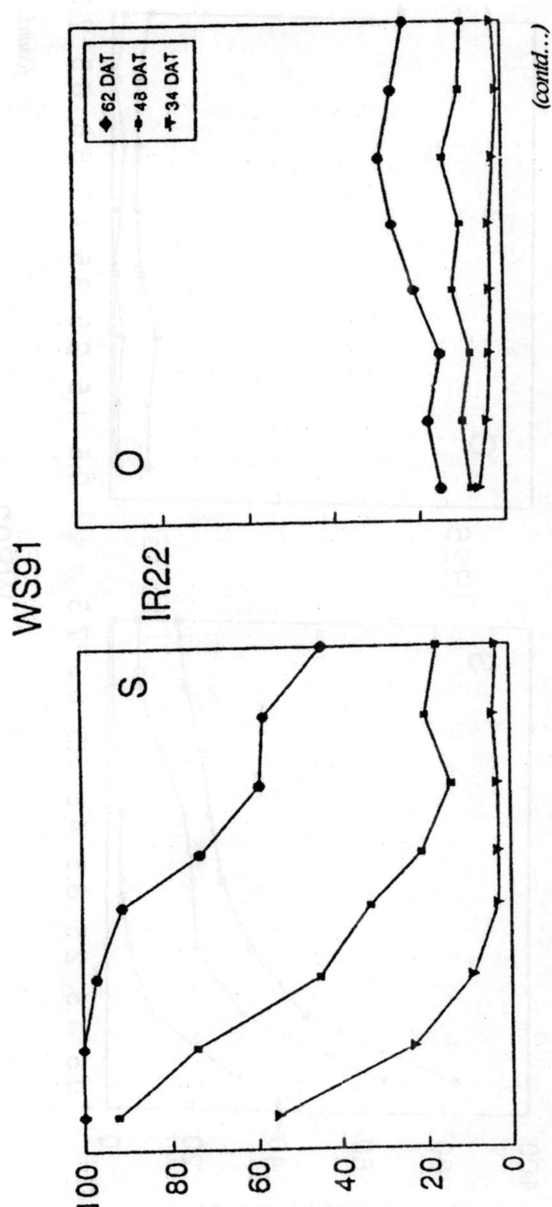

(contd...)

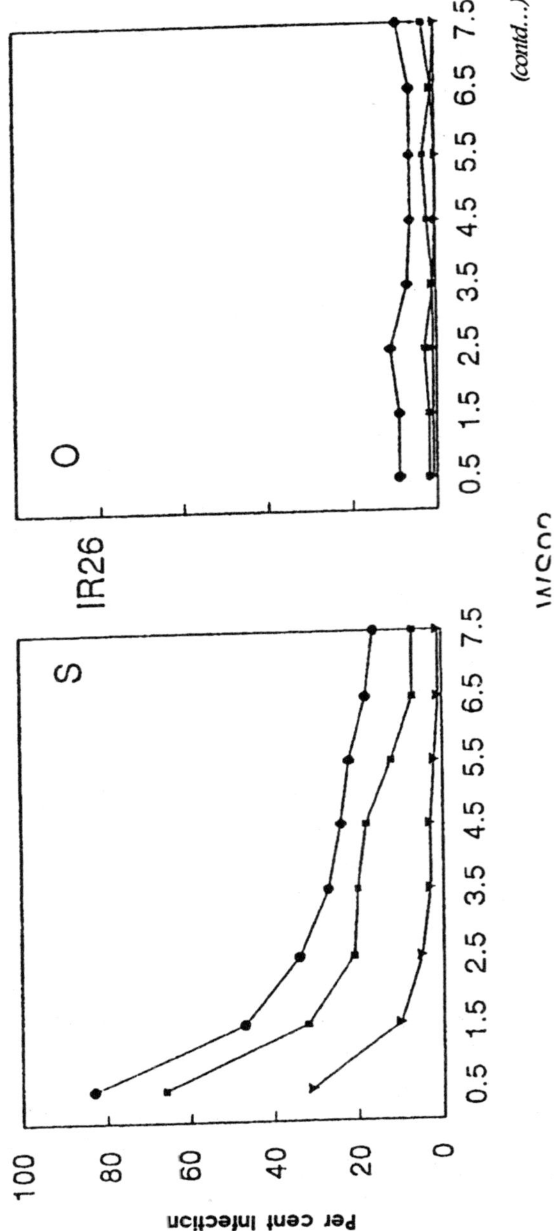
IR26
S
O
Per cent infection
100
80
60
40
20
0
0.5 1.5 2.5 3.5 4.5 5.5 6.5 7.5
0.5 1.5 2.5 3.5 4.5 5.5 6.5 7.5

(contd...)

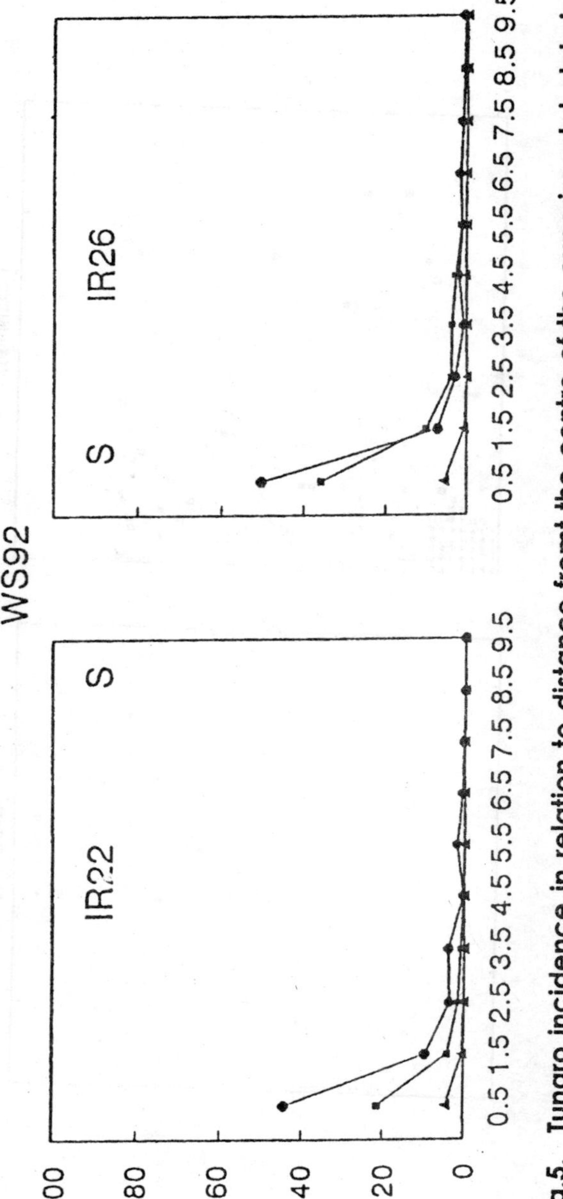

Fig.5. Tungro incidence in relation to distance fromt the centre of the experimental plots in IR22 and IR26 with (S) and without (O) introduced sources of inoculum in 1991 and 1992 wet season (WS).

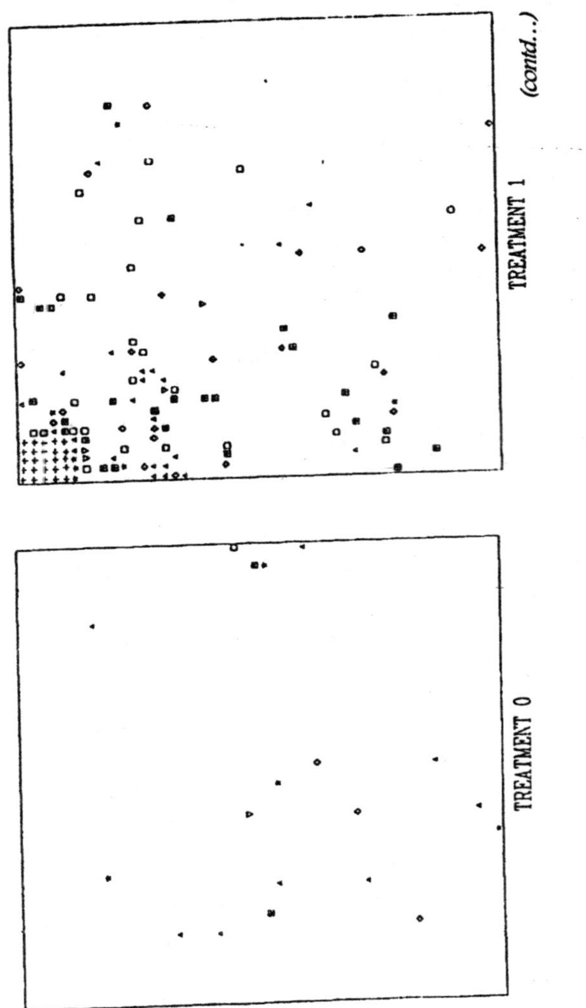

(contd...)

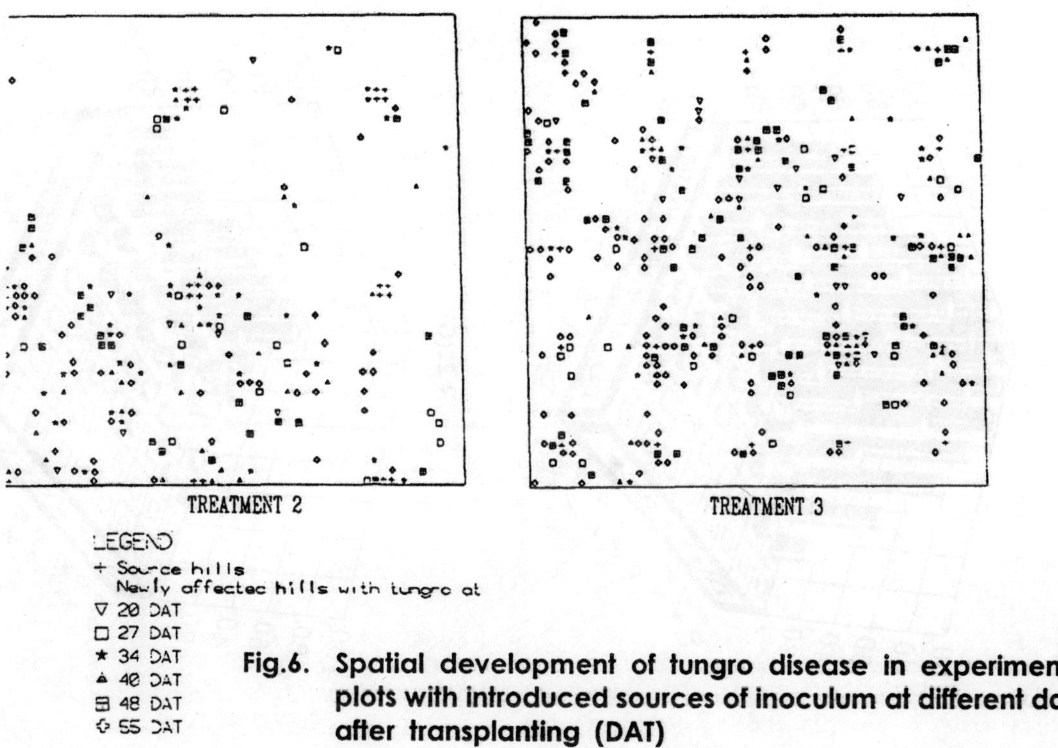

Fig.6. Spatial development of tungro disease in experimental plots with introduced sources of inoculum at different days after transplanting (DAT)

IR 22(S)

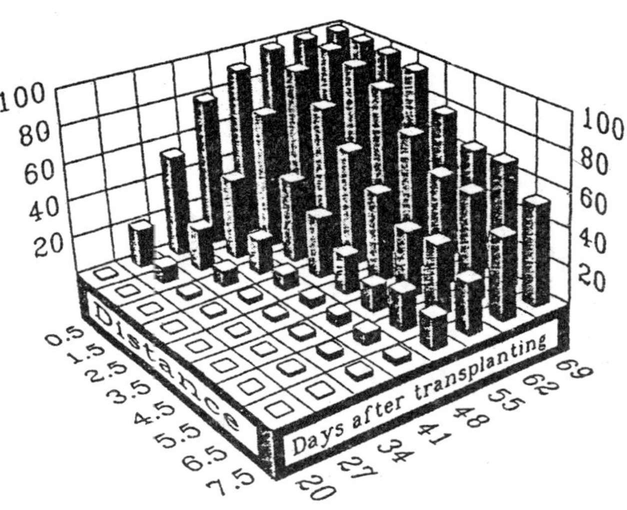

IR 22(O)

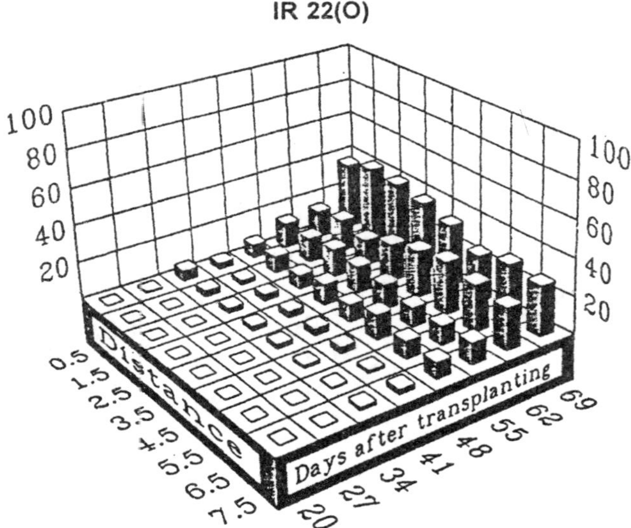

Fig.7. Spatio-temporal dynamics of tungro disease in cultivar IR22 with (S) and without (O) inoculum source

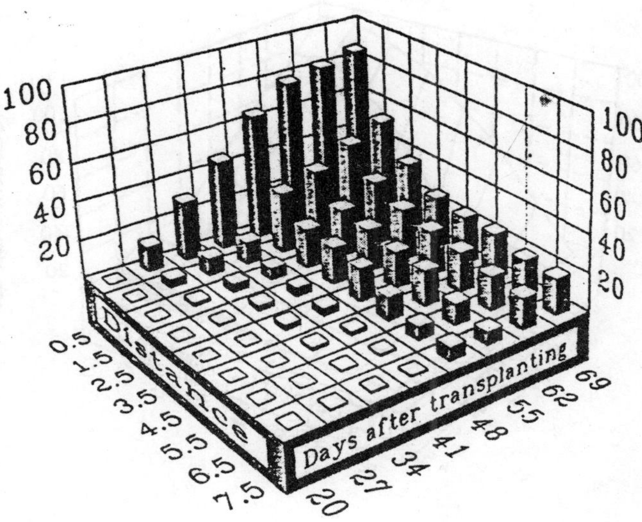

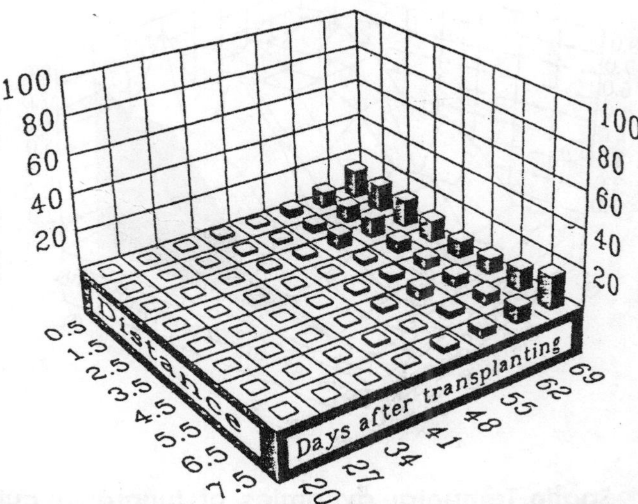

Fig.8. Spatio-temporal dynamics of tungro disease in cultivar IR26 with (S) and without (O) inoculum source

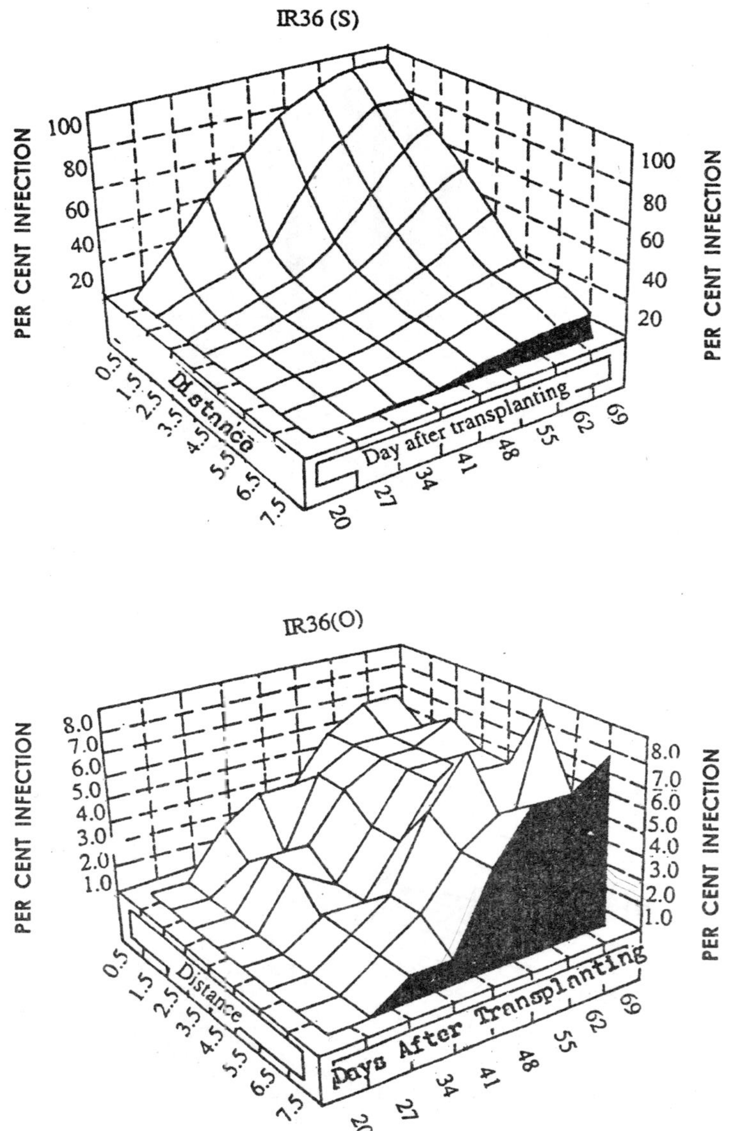

Fig.9. Spatio-Temporal dynamics of tungro in cultivar IR36 with (S) and without (O) inoculum sources

IR72 (S)

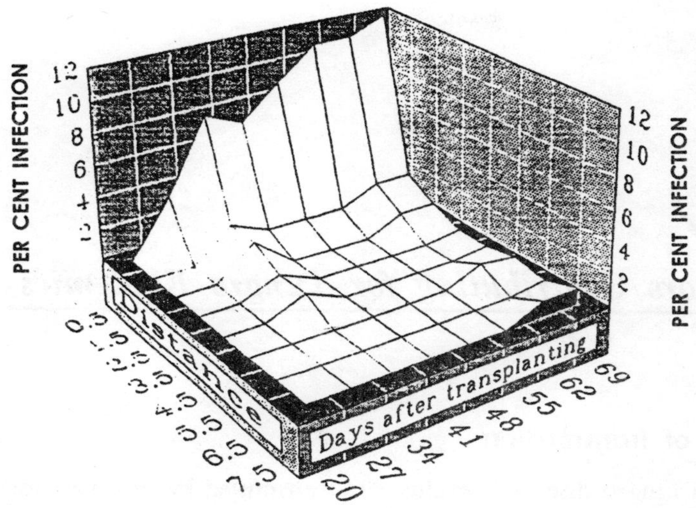

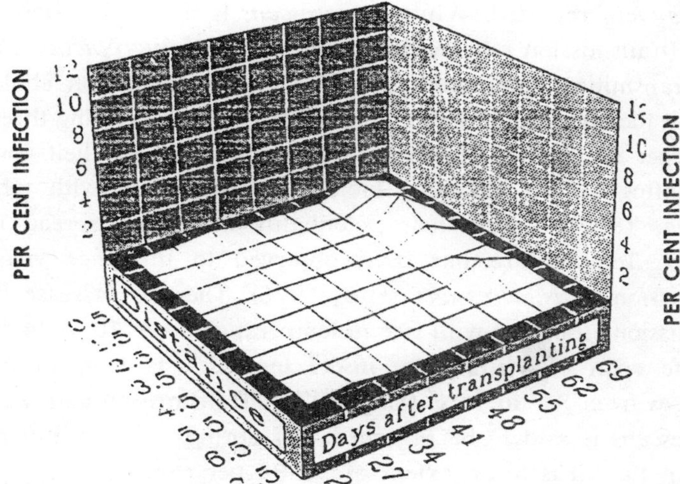

Fig. 10. Spatio-temporal dynamics of tungro in cultivar IR72 (Wet season 1991)

8

Factors Contributing for Tungro Epidemics

Ease of Transmission

Rice tungro disease is exclusively transmitted by insect vectors (rice green leafhoppers). *N. virescens* was first demonstrated to be the vector of tungro by Rivera and Ou (1965). Subsequently *N. nigropictus*, *N. cincticeps*, *N. malayanus*, *N. parvus* and *Recelia dorsalis* were reported. While *N. virescens* is the most efficient vector (transmission rate reaching upto 83%), *N. nigropictus* is a poor transmitter (miximum upto 27% of the population could transmit the virus) Ling, 1970; Singh, 1971). The remaining three leafhopper species are inefficient vectors, because of their low transmitting ability and poor biological relationship with rice (Sogawa, 1976). Hence their potentiality in disease spread is doubted. Temperature has been observed to influence virus transmission in *N. virescens*. (Ling, 1976). There is increase in transmission efficiency with rice in temperature from 10 to 34°C. The life span of viruliferous insect increases as temperature decreases from 34° to 13°C (Ling, 1976). Transmission ability of *N. virescens* is said to be inherited. Off springs of transmitters transmit the virus to an extent of 90-94 per cent.

Female *N. virescens* transmits tungro viruses more efficiently than male (Lamey *et al.*, 1967; John, 1968). Numphs of all growth stages transmit tungro viruses like adults; but the third, fourth and fifth instar nymphs are more efficient than first and second instars (Carbonell and Ling 1971; Hino *et al.*, 1974; Anjaneyulu, 1975a). Though nymphs are less efficient in virus transmission (Ling, 1975a) than adults, they play very active role in transmitting disease within a field secondarily by jumping from plant to plant. Though disease spread within a field is carried out both by nymphs and adults green leafhoppers, inoculum spreads from field to field (among fields) in a locality and among localities through migration/ movement of viruliferous adult hoppers. Thus in epidemic years the rate and ease of transmission might be playing an important role in disease spread. Besides transmission rate virus-vector relationship is also responsible for quick transmission and spread of the disease.

Unusual Virus-Vector-Hose Relationship

Generally many leafhopper borne viruses and almost all mycoplasma sp. are propagative in their respective vectors. The virus multiplies both in the plant host as well as in the insect vector, though the tissues of multiplication are different. It is interesting to note that the virus causes much damage in the hose plant but it is passive in the insect vector.

In propagative viruses, it requires about one to two weeks time before becoming infectious. During this period, the virus multiplies, passes through a cycle and reaches the mouthparts of the vector through the salivary gland to cause infection. Its infectivity is not affected by moulting and it remains infective throughout its life (Kiritani, 1981). In certain cases the virus passes through the egg of the insect vector and spreads for several generations, thus the vectors retain their infectivity through successive generations without access to a diseased rice plant as is the case with rice dwarf virus transmitted by *N. cincticeps*, *N. nigopictus*, *N. virescens* etc (Hibino 1989).

In rice tungro viruses, there in no evidence of virus multiplying in the vector. In the host plant both the viruses (RTSV and RTBV) multiply independently. In leafhopper-susceptible cultivars the leafhopper feeds mainly from the phloem and transmits both the viruses (RTSV + RTBV), while in resistant cultivars it feeds on xylem and during feeding it transmits RTSV also (Heinrichs and Rapusus, 1984). When the viruses are acquired by the insect vector, they are a desorbed and get distributed on the surface of the stylet. The viruses do not enter into the body of the insect vector and there is no incubation period in the vector. Within few minutes of virus acquisition (Rivera and Ou, 1965; Ling, 1969) or even as low as 30 seconds of feeding (Dahal *et al.*, 1990b) the vector can infect another host plant. Even a single probing by an infective insect is said to cause disease in a plant (Ling, 1968). The length of acquisition access time affects the percentage of infective insects and they transmit the disease soon after acquisition access feeding and every day until they lose infectivity of course the infectivity is lost by moulting. Pre-acquisition starving improves the transmission efficiency of the vector but the post-acquisition starving reduces it (Ling, 1966; Singh, 1971). Once acquired the vector can retain and transmit the viruses for about 5 days in tropical temperature, but at 13°c transmitting time increases upto 22 days (Ling and Tiongco, 1977). Until this the virus was considered to be non-persistent and subsequently Ling and Tiongco (1977) proposed a new term 'transitory' for it. However this type of virus-vector relationship is popularly called as semi-persistent, and it is very unusual among leafhopper transmitted viruses.

The retention period in the vector depends upon the length of acquisition access. The retention period is about 2 days at 4-16 hr, 3 days at 24-72 hrs and four days at 96-144 hrs of acquisition access time. Among the nymphs the maximum retention period is about two days for the first-3rd instars and three days for 4th and 5th instars (Rao, 1977). The retention period varies with vector species, for example it is for three days for *N. nigropictus* and four days for *R. dosalis* (Rivera *et al.*, 1969). Further, it has been observed (*Khan et al.*, 1991) that both RTSV and RTBV are retained for four days in *N. virescens*, five days in *N. nigropictus* after an acquision feeding of about one day.

Infective insects usually transmit the disease soon after acquision access feeding and every day units they loose infectivity, unless given access to another virus source. There is decrease in infectivity with time following acquisition feeding. Some insects transmit the virus throughout their lives if given a requisition feeding every day (Ling, 1966). Insects lay eggs in the wings of leaf at a mass of single rows reading upto 365 in number and the oviposition period is about 10-14 days with a pre-oviposition time of 7-8 days. The average life span of insects is about 25-26 days and the females survive longer than the males. Because of short life span with a very short pre-oviposition period insects build up at a fast rate and help in the transmission of viruses throughout the cropping season. Interestingly the green leafhopper has about 3 to 4 generations per crop period with the second having the highest density.

The virus being systemic is distributed throughout the host tissue. The insects acquire viruses more easily from young plants than old. The incubation (latent) period in the host plant is five to six days and it increases with increase in host age. The ability of tungro viruses to persist in the vector for few days or about a week and the short feeding periods needed for acquisition and transmission create the potential for rapid disease spread in the field. In these viruses, complete infection of the plots is a rare and occasional phenomenon. Moreover the short latent period in the host especially young rice plants and consequent occurrence of several disease cycles in a cropping season helps the disease to cover vast areas and attain epidemic level.

Quick movement of the vector *N. virescens* in the field

Movement of the insect vector is one of the important factors responsible for epidemic outbreak. Of different insect vectors, white flies move fast and often cause epidemics. Following white flies are the green leafhoppers.

N. virescens is an active vector of tungro viruses. Numphs jump either directly from plant to plant or from plant to water and again to plant and spread much the disease (Anjaneyulu, 1975a) within a field. It has been observed that much of the secondary spread that occurs in small circular patches within a plot is carried out primarily by nymphs (A. Anjaneyulu- personal observation). This is of much importance in areas where rice is intensively cultivated throughout the year as a monoculture.

Occurrence of circular patches is explained by the fact that the source is initiated probably with the arrival of a flying adult infective hoppers that sparks the infection. Further spread occurs by the local nymphs jumping from plant to plant. It is possible to control the secondary spread by killing numphs through systemic insecticides like carbofuran or contract insecticides like cypermethrin.

Adults of *N. virescens* are considered to be poor fliers (Chancellor and Cook, 1995), Kondaiah *et al.* (1976) from epidemiological studies on tungro and Bottenberg and Litsinger (1989) from their movement studies of *N. virescens* through use of fluorescent dyes have observed localised movement of the insect over a distance of few meters. This local movement helps disease spread within a field or of few fields within a locality. This is of much importance in areas where rice is intensively cultivated throughout the year as a monoculture.

Migration of *N. virescens*

Nephotettix spp. exhibits a marked bimodal crepuscular pattern of flight with a larger peak occurring in the evening (Perfect and Cook, 1982). Though flight activity continues throughout the night (Chancellor *et al.*, 1996a) evidence from rader studies in the Philippines shows it to be mostly confined to periods of about 30 min at dusk and dawn (Riley *et al.*, 1987) suggesting that the insect could disperse for few kilometers. Net trap catches on ships in South China sea which confirmed the long distance migration of the planthopper *Nilaparvata* lugens (stoal) and *Sogatella furcifera* (Horvath), included only a few specimens of *N. virescens.* A few specimens of *N. virescens* were also collected on interisland ships up to 30 km from possible sources in the Philippines (Saxena and Justo, 1986). Riley *et al.* (1987) from their studies on migration of rice plant and leafhoppers in the Philippines using rader technique have concluded that *N. virescens* regularly undertakes migratory flights of about 5.30 km.

Using tethered flights techniques, Cooter and Winder (1996) have studied flight activity and dispersal characteristics of *N. virescens.* They observed that most of the hoppers make short flight while few fly for long distance which has an important implication in tungro virus inoculum spread between fields. Besides seasonal and meteorological factors age of the host, age and se of insects influence their flight activities. The hoppers undergo long flights when exposed senescening rice plants and diseased plants and moderately resistant plants which has important implications for disease spread. Besides biological factors such as nature of host plant, natural enemies, velocity and direction of wind also (Mukhopadhyay, 1980) influence movement and dispersal.

In large irrigated areas where rice is grown as monoculture round the year, the selective advantage of long-distance migration is greatly reduced. However in single cropping areas where host plants are not available to support population throughout the year, selection pressure might be helping in the development of populations having a high proportion of flight active individuals.

Besides host plant inducing movement and migration, it has commonly been observed that females of *N. virescens* show higher response to lunar periodicity than males of the species. Further females make more flight after getting sexually matured. Thus maturing rice crop and female hoppers are not only responsible for the mass migration of insects but they probably play some indirect role in the initiation of the disease.

Population dynamics of the Vector

Tungro being a vector borne virus disease, availability of vectors (*Nephotettix spp*), their population size and time of occurrence besides other factors such as transmission efficiency play very important role in disease outbreak. In warm and humid tropics, vector population fluctuates depending on the availability of host plants, cropping pattern, environmental conditions, natural enemies etc. *N. virescens*, the prime vector for tungro viruses prefers rice plant but in its absence it moves to weeds and grasses for hibernation or aestivation.

Annual Population Fluctuation

In tropical Asian countries where one main rice crop is grown in the year (rainy season), *N. virescens* population shows a distinct annual peak. In irrigated areas with two or more rice crops, two population peaks are observed with the major peak during rainy season. In India, a single peak is usually observed in the middle of the kharif season but little variation is seen from state to state. The leafhopper population reaches a peak in July-August in Assam, Manipur, and Tripura; August-September in West Bengal and September-October in Orissa, Bihar, Uttar Pradesh, Madhya Pradesh and Andhra Pradesh (Anjaneyulu, 1974; Gupta *et al.*, 1989). In Thailand where a single rice crop is usually grown in rainy season (June/July-November/December) the population reaches a peak during July-September (Inoue *et.al.*, 1975). In the Philippines where double cropping in commonly practised, two peaks are observed, the main peak during September-October and the minor peak during January-February (IRRI, 1973). In Indonesia where double cropping is seen, two peaks (one in wet and other in dry season) of green leafhoppers are seen.

Seasonal Population Development

In single cropping areas where rice is grown during rainy season, green leafhoppers migrate from regenerated stubbles (if any), wild rices, weeds into the seedbeds from where they move into the rice fields. In areas where overlapping cropping or asynchronous planting in seen, insect vectors migrate from established crops in the vicinity (Chancellor and Cook, 1995) whereas in double cropping or synchronous planting areas where there is an interval between two cropping seasons (fallow period) the immigration most likely occurs from regenerated rice stubbles into the seedbeds.

In the Philippines where two or three rice crops are grown in a year, the leafhoppers are seen round the year. In West Bengal (India) the insects appear towards the end of dry season (March-April), gradually increase during wet season and reach a peak during September-October. The population declines slowly during winter season (November-February) Mukhopadhyay and Chowdhuary, 1973; Mukhopadhyay, 1986). In Bangaladesh (Alam and Islam, 1959) *N. virescens* multiplies in aus crop (April-August) and reaches a peak towards the end of aus crop and begining of aman crop (July-August). In double cropping areas of Malaysia, *N. virescens* increases rapidly during May-June, reaches a peak during July, and declines sharply thereafter (Lim, 1969). Large number of *N. virescens* also appear in the seedbeds and migrate to late planted crops (Bottenberg *et.al.*, 1990). Green leafhopper (GLH) population dynamics at two sites, one each in Indonesia and Malaysia has been shown in Fig. 11.

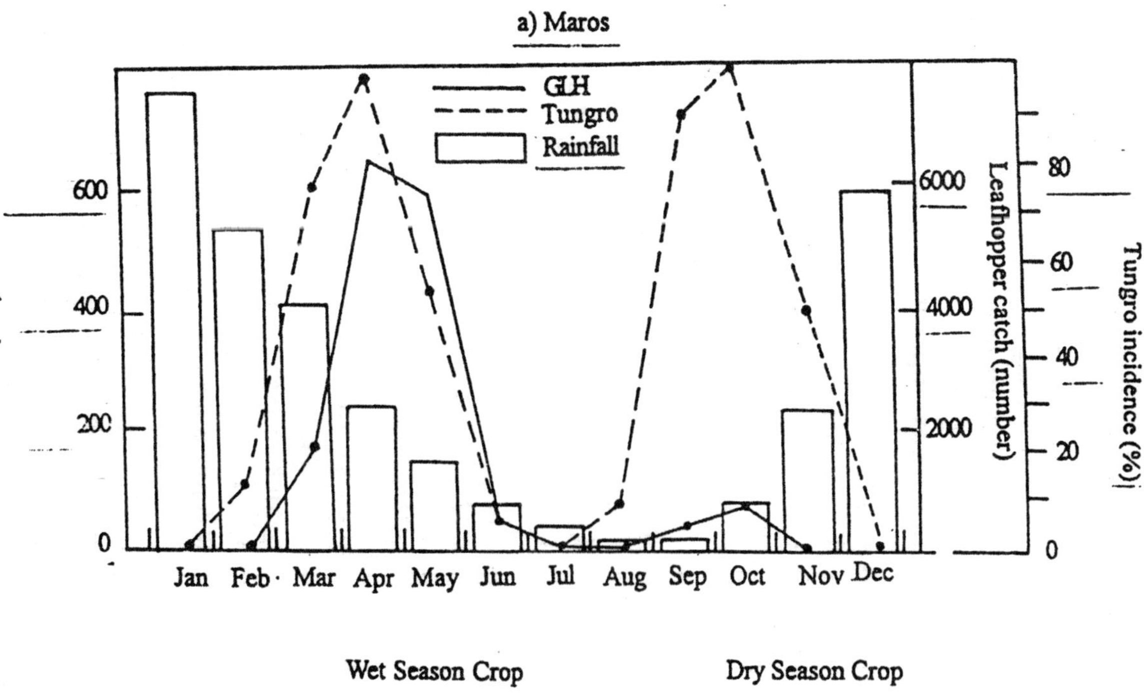
a) Maros
GLH
Tungro
Rainfall
Rainfall (mm)
600
400
200
0
Leafhopper catch (number)
6000
4000
2000
Tungro incidence (%)
80
60
40
20
0
Jan
Feb
Mar
Apr
May
Jun
Jul
Aug
Sep
Oct
Nov
Dec
Wet Season Crop
Dry Season Crop

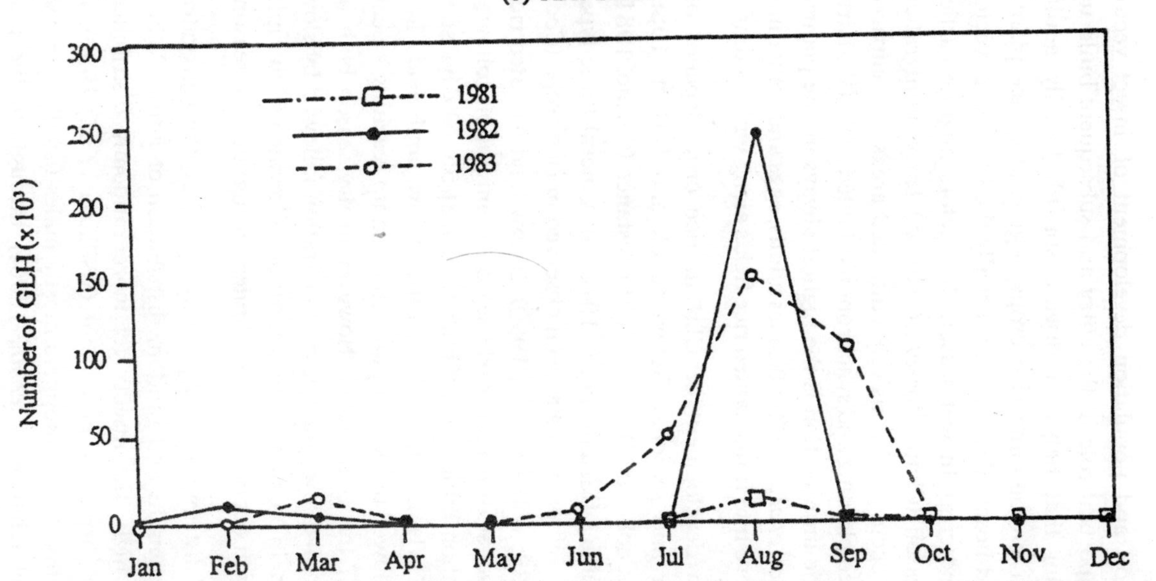

Fig.11. Green leafhopper (GLF) catches in light traps for wet and dry seasons, at a) Maros, Indonesia (1997-81) and (b) Alor Setar Malaysia (1981-83) (Chancellor and Thresh, 1997 with modification).

Immigration and Population Development in Rice Crops

Immigration and population development of insect vector plays a very important role in the entry and subsequent build up of the inoculum that helps in disease spread. Usually adults immigrate from surrounding rice crops, regenerated rice plants, stubbles or weed hosts. *N virescens* is usually the first insect vector to appear on the crop. In wet season the vectors migrate earlier than the dry season (Chancellor *et al.* 1996c.) Early immigration of *N. virescens* occurs in intensively cultivated areas as compared rainfed, staggered planting and synchronous planted areas (Widiarta *et al.*, 1990). The immigration is the highest always in late planted crops (Chancellor *et al.*, 1996b.) Besides adults, eggs and sometimes nymphs also move into transplanted rice fields along with seedings..

Population development of GLH in rice crop depends on cropping patter and seasonal factors (Cook and Perfect. 1989; Chancellor and Cook, 1995) besides host resistance (Carino, 1981) and natural enemies (Kiritani *et al.*, 1970). In general three types of population development have been observed in rice crops (Cook and Perfect, 1989; Widiart *et al.*, 1990). In low land irrigated rice fields, during wet season high early season immigration followed by rapid population build up reaching a peak at 55-70 days after transplanting (DAT) is observed. In rainfed low land fields low immigration followed by slow population build up reaching a peak at 102-110 DAT has been noted. However in dry season both in irrigated and rainfed low lands, low immigration followed by slow population build up is common for some time following which rapid population growth occurs and it continues throughout the season.

The spatial distribution of *Nephotettix spp.* shows an aggregated distribution of nymphs and random distribution of adults. When immigration is high, the insects get more randomly distributed facilitating their dispersive movement (Widiarta *et al.*, 1990). It has been noted that the leafhoppers mostly males usually disperse at cannopy level (chancellor *et al.*, 1996c) throughout the plot creating the potential for much secondary plant to plant spread of tungro disease.

Usually it has been noted by several researches that occurrence of tungro epidemic covering vast areas usually concicides with the peak period of the green leafhopper population growth. Probably once the inoculum is available through immigrant hoppers coming from the surroundings or through the infected seedlings coming from the seedbeds, the population build up enhance the spread of the inoculum and the disease through dispersal if the correct (young) growth stage of the rice plant and susceptible variety are available. The population dynamic of GLH always coincides with the weather/climatic condition of the area that in turn favours cultivation and active growth stage of rice crop.

Vector's Attraction to light

The green leafhoppers have definite attraction to light. Besides light colour, light intensity has an important role as population size of insect vectors attracted towards light depends on light intensity. (Perfect and Cook, 1983). At Cuttack (Orissa, India) if has been observed by the authors that during Dasahara-Diwali period (October-November) leafhoppers migrate in huge numbers towards light and cause a lot of inconvenience to people. During the same time leafhoppers are seen in huge numbers on the light points in the middle of Calcutta city, about 20 to 30 km. away from the nearest rice fields. It is a subject of study whether the hoppers fly this long distance at a stretch or by several hoppings.

Leafhopper attraction to light is probably considered to be one of the major factors for the outbreak of tungro epidemic. In 1990 the entire east coast of India experienced a severe tungro epidemic. In this area there are several railway tracks and one can see numerous leafhoppers moving with the train in the night being attracted toward its headlight. The authors suspect that viruliferous hoppers while moving along with train light probably migrate to rice fields causing infection. Train stops at stations on its way, some hoppers fly always from the headlight to nearby rice fields and other leafhoppers come in.

Since leafhoppers move along with the trains, more diseased plants are expected to be seen in plots along the side of train tracks. But this is not the situation always. During 1990 epidemics the second author (AA) has observed diseased plants as far as 100 km away from train lines. This could have been due to movement of viruliferous hoppers along with head lights of moving vehicles (trucks/buses) running on National/State highways. The author while travelling during day time in train from Vijayawada to Cuttack via Visakhapatanam (a distance of about 800 KM in the Madras Howrah track) observed severe tungro disease symptoms in rice fields along both sides of the track. This needs further investigation and analysis as probably it is one of the factors that help in the spread of the inoculum (in terms of availability of viruliferous hoppers) for disease spread in a rice growing region. In all the years tungro outbreak does not occur, probably through hoppers become available and rice is regularly cultivated, inoculum becomes the limiting factor.

In the same year the second author (AA) while visiting certain villages in the district of East Godavari and Srikakulum (in the state of Andhra Pradesh) observed severe tungro infection in areas with intensive cultivation of high yielding IR64 variety as well as in areas undertaking overlapping cropping practices. The fields were severely infected and farmers were equally worried for straw, like grains as the former is an important feed for their cattles.

It was noted that farmers of these localities have a craze for the rice cultivar IR64 not only for its high yielding nature but the farmers easily grow two crops in a single year from the dame field. There is a slight overlapping of two crops, while harvesting the first crop they raise nursery bed for the second crop. IR64 being a susceptible variety, its intensive cultivation has made the variety susceptible with the available of virus inoculum, the disease spreads very fast. Under this situation probably light sources from train/vehicles might be helping in the easy dispersal of inoculum.

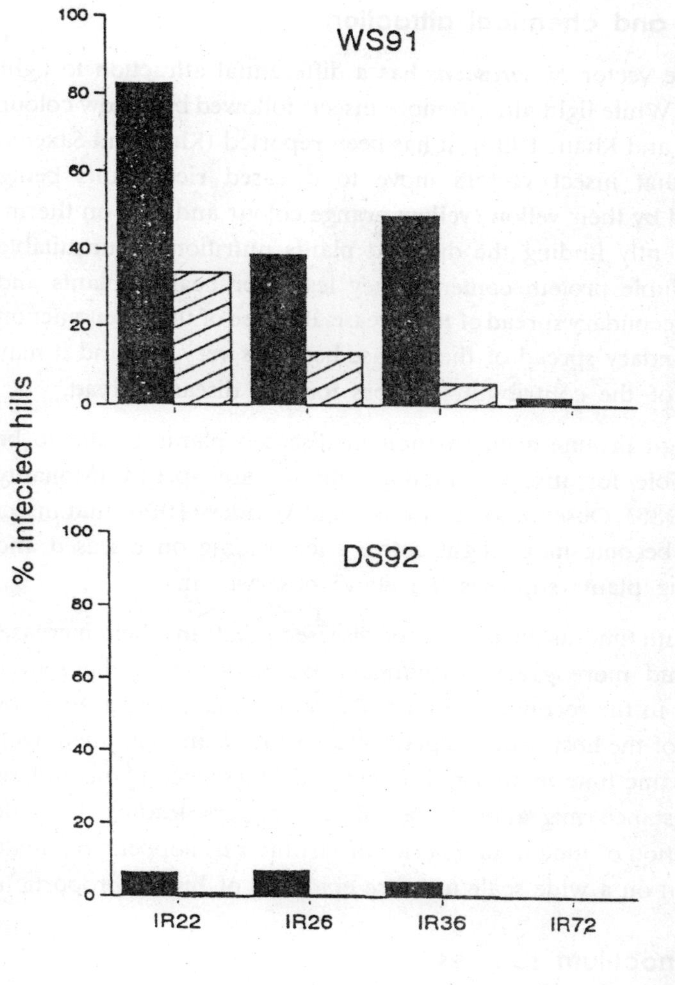

Fig.12. Tungro disease incidence in four rice varieties wit and without introduced sources of inoculum in 1991 wet season (WS91) and 1992 dry season (DS 92) experiments at 69 days after transplanting (DAT)

Colour and chemical attraction

The vector *N. virescens* has a differential attraction to light colours. White light attracts more insects followed by yellow colour (Abenes and Khan, 1990). It has been reported (Khan and Saxena, 1985) that insect vectors move to diseased rice plants being attracted by their yellow/yellow orange colour and feed on therm. Subsequently finding the diseased plants nutritionally unsuitable (low soluble protein content), they leave for healthy plants and help in secondary spread of the disease. Because of this phenomenon the secondary spread of the disease becomes very fast and it may be one of the contributing factors for fast disease spread.

High proline accumulation in diseased plants is said to be responsible for insect attraction and disease spread (Mohanty *et al.*, 1983). Observation by Cooter and Winder (1996) that insect vectors become more flight active after feeding on diseased and senescing plants supports the above observation.

With time, as the number of diseased plants in a field increase, more and more green leafhoppers are attracted to the source and aid in the secondary spread. Of course this is constrained by the age of the host plant (as plants become resistant with their age). At the same time maturing diseased and senescence plants induce long distance migration/dispersal of hoppers leading to wide distribution of inoculum in terms of viruliferous hoppers to initiate infection on a wide scale to cause epidemic of higher proportion.

Virus inoculum sources

Tungro disease spread within and among rice fields depends on the availability of inoculum sources (Fig.12), their amount and disposition. The ability of any plant species to act as virus source depends not only on the concentration of virues in its tissues but also on the vector's ability to acquire it. Rice cultivars their stubbles, volunteer rice plants, wild rices, weeds and grasses have been recorded as potential inoculum sources for tungro disease spread.

Rice Cultivars, Stubbles and Volunteer Rice Plants

As potential inoculum sources for tungro viruses rice cultivars do vary. Generally susceptible cultivars are more potential in terms of producing higher percentage of infective insects and longer retention period of the viruses than resistant cultivars. Cultivars like TN1, IR22, IR 24, Padma, Bala, Krishna etc have been observed to act as good inoculum sources (Ling 1975b; Hasanuddin and Ling, 1980; Anjaneyulu *et al.*, 1982b) as compared to IR20, IR26, IR30, Ratna, Annapurna, and PUsa 2-21, CR 138-802-10 etc. Age of the rice plant irrespective of the nature of the variety has a strong impact on the variety acting as tungro sources. Usually young seedlings act as better source of infection than old plants. Certain cultivars like Jaya recovers three to four weeks after infection and look green. Such varieties also carry tungro viruses. Further it has been noted (IRRI, 1987) that virus concentration varies with the leaf. Usually second and third youngest leaves carry higher amount of virus than younger or older leaves.

Besides infected rice plants, their stubbles have been observed to carry tungro viruses. Stubbles of susceptible plants such as IR8, IR22, TN1, Jaya, Padma, Bala, Krishna and Pankaj act as potential reservoir for viruses between two crops and help in disease spread. (Anjaneyulu 1975b; Anjaneyulu *et al.*, 1982b). Stubbles of resistant varieties like IR20, IR26, IR30 Annapurna, Ratna, CR 138-994-27 are poor sources of viruses. The stubbles of kharif crop carry more viruses than those of boro crop (Tarafder and Mukhopadhyay, 1979a). Further it was noted that stables of Jaya and Pankaj carry viruses for about 60 days whereas for Ratna it is only for 30 days. Besides stubbles, volunteer rice plants (Tiongco *et al.*, 1992) coming out from seeds spilled during harvest and also ratoons (Ali and Miah, 1990) harbour viruses as well as vectors and act as potential sources of infection in the field especially in asynchronously planted areas. When stubbles are puddled in the soil or are chopped off they fail to act as sources of infection may be because of failure in feeding by the leafhoppers (Mukhopadhyay,

1980). Interestingly Mishra *et al.* (1983) found that infected rice leaves even after 28 months of storage carry viruses and act as source for initiating infection. The importance of inoculum source on tungro epidemiology has been shown in Fig. 12.

Wild Rices

About 12 species of wild rice such as *Oryza australiensis*, *O. barthii*, *O.officinalis*, *O. redleyi*, *O. rufipogon*, *O. fatua*, *O. glaberrima*, *O. nivera*, *O. perennis*, *O. brachyantha*, *O. eichengeri*, *O. punctata* etc. have been found susceptible to tungro infection in artificial inoculation tests in different Asian countries (Ting and Ong, 1974; Hino *et al.*, 1974; Anjaneyulu *et al.*, 1981; 1982b). It has been possible to recover viruses from *O. barthii*, *O. glaberrima*, *O nivera* and *O. sativa* through feeding by non-viruliferous green leafhoppers. Though with ELISA, infection with RTSV and RTBV has been ascertained in 10 wild species (Parejarearn *et al.*, 1990) none of these plants have been observed as natural sources of infection in the field.

Since wild rices have not been observed acting as sources of infection in the field, they are relatively resistant to virus infection. However it is possible that under high insect and disease pressure they might be acting as sources of infection and playing some role in initiating the disease. Of course it needs a clinical scruting.

Weeds and Grasses

Many weeds and wild plants such as *Eleusine indica*, *Leersia hexandra*, *Rottboellica compressa*, *Cyanodon dactylon*, *E. indica*, *Hamarthria compressa*, *Polypogon monspeliensis*, *Echinochloa colonum*, *E. crusgalli*, *Dactyloctenium aegypticum*, *Eragrostis tenella*, *Ischinum rogosum*, *Paspalum scrobiculatum*, *Setaria glauca*, *Sorghum vulgare*, *Bothriochloa odorata*, *Brachiaria reptans*, *Digitaria adescendens* and *Pennisetum typhoides* have been found as carriers of tungro viruses through artificial inoculation tests (Wathanakul 1964; Raychaudhuri *et al.*, 1971; Mishra *et al.*, 1973; Mukhopadhyay *et al.*, 1978). However many of these wild hosts do not show clear symptoms and attempts have failed to recover the viruses from them. Very few species have been found infected in the field. Few species such as *Eleusine indica, L. hexandra, Rottboellica compressa, Cyanodon dactylon* have been found infected in the field. *E. colonum* has been found to be a good source of virus infection. This species retains the virus for four months in April inoculation whereas it can carry virus for three months when inoculated in November.

Using serological techniques presence of RTBV and RTSV have been found in *Hydrolia zeylanica, C. rotundus, Fimbristylis miliacea, L. hexandra, E. indica, E. crusgalli, E. glabrescens, L. chinensis* and *I. rogosum* (Mohanty *et al.*, 1987; Anjaneyulu *et al.*, 1988; Khan *et al.*, 1991; Mallick *et al.*, 199) infected naturally or through artificial means. However with *N. nigropictus* as vector, viruses have been transmitted from *E. crusgalli, P. repens* and *c. rotundus* to rice seedlings (Khan *et al.*, 1991).

A critical observation of tungro virus sources reveals that in irrigated asynchronously planted areas, tungro viruses might be carried over through standing rice crop, nursery beds, stubbles, ratoons or volunteer seedlings from one rice crop to the other (Anjaneyulu *et al.*, 1982b; Rao and Hasanduddin, 1991) as rice is the preferred host both for vector and the viruses. In rainfed single crop or synchronously planted areas, because of a crop free period and absence of rice the carryover is possibly either through long distance migration of viruliferous hoppers or through weeds and wild rices growing in the vicinity or both. Since wild rices have not been observed as natural sources for tungro viruses, possibly some weeds might be acting as sources for initiation of the disease. The weeds might be acting as alternate hosts and reservoir for the viruses in the rice fields.

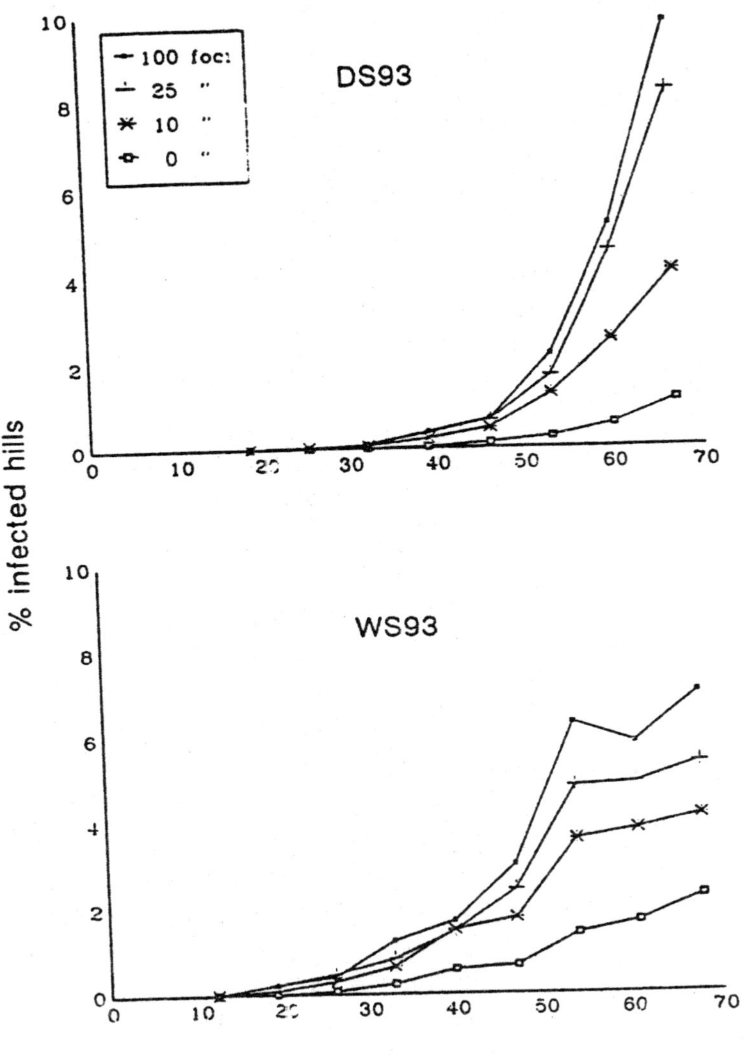

Fig. 13 Progress curves of rice tungro disease in IR22 rice with different foci of infection during dry and wet seasons of 1993

Experimental transmission of RTSV and RTBV from rice to certain weeds through *N. nigropictus* and back (Khan *et al.*, 1991) adds evidence to it and this may be responsible for triggering the epidemics. Possibly within or towards the end of the season insect vector, *N. nigropictus* carries virus from rice plant to weeds in which they survive during crop free period. Again with the onset of cropping season, the insect *N. nigropictus* again transfer the viruses from weeds to rice plant. Once the virus is available in rice plants further spread within the field occurs through *N. virescens.* It leads to outbreak of epidemics. However its needs further investigation.

Potency and Disposition of Inoculum Sources

Nature of inoculum sources (rice/weed) and its effect on disease spread has been discussed earlier. Availability of inoculum sources within the plot/field in the form of diseased hills or immigrant infective vectors has a profound impact on disease spread as most of the disease spread occurs secondarily because of plant to plant spread by leafhoppers that acquire tungro viruses from sources within the plots (Satapathy *et al.*, 1996). Primary inoculum sources play very important role in the disease initiation at the early stages of crop growth when plants are in most vulnerable stage. Once infection starts, subsequently much spread occurs because of secondary spread from the existing sources within the plot.

Moreover the disease spread increases with increase in inoculum size (Table 2) (Shukla and Anjaneyulu, 1982a; Chowdhuary *et al.*, 1994 Satapathy *et al.*, 1996) as high density of inoculum increases the probability of visit of large number of green leafhopper vectors to the source plants. They feed, acquire the virus and help in further spread of the disease (Khan and Sexena, 1985). Moreover with increase in source size chances of vectors becoming viruliferous also becomes high. Further dispersed distribution of inoculum sources provide large number of foci (Fig. 13) for secondary spread of the disease (Table 3). In the fields the inoculum is likely to arrive randomly either through the infected seedlings coming from the seedbeds or through immigrant viruliferous hoppers landing irregularly in the plot. This leads to presence of more number of foci for disease initiation that enhances the changes of secondary spread. For example, the seedlings infected in the seed bed are planted, the disease is initiated early. Eggs led by green leafhoppers in the leaf sheath of seedlings transplanted in the field hatch soonafter transplantation and help in the disease spread. With time when vector population size increases disease spreads very fast. In both cases disease spreads very fast and reaches a high level if it coincides with young growth stage of the rice crop and abundant green leafhopper population leading to breakdown of epidemics.

Rapid multiplication of tungro viruses and their spread in the host plant

In leafhopper transmitted viruses usually it takes about 2 to 3 weeks time after infection for the expression of symptoms in the host plants and they become infectious. But interestingly infection with tungro viruses, the rice plants at young age take only 4-5 days to become infectious and 6-7 days for clear expression of symptoms. This is due to fast multiplication of both (RTSV+RTBV) viruses in host tissues. The rice plants become early and more infectious during wet season because of luxuriant growth of plants and early symption expression in them. Moreover the viruses being systemic, they are acquired from any tissues of the host including semi dried leaves. All these factors help in quick spread of the disease in the field (Rao, 1977).

Table 2 : Area under disease progress curves (AUDPC) and infection rate in relation to source size in IR22 and IR36 (1992 WS) and IR22 (1993 DS) (Satapathy *et al.*, 1996).

Source size (no. hills)	*AUDPC*				*Infection rate (r°)*		
	1992 WS			*1993 DS*	*1992 WS*		*1993 DS*
	IR22	*IR36*	*Diff*	*IR22*	*IR22*	*IR36*	*IR22*
0	553ab	135a	418**	6a	0.129	0.098	0.07
1	516a	160a	356**	31a	0.154	0.115	0.09
5	495a	289a	206ns	39a	0.167	0.112	0.09
25	801b	360a	441**	129b	0.129	0.103	0.10
125	1230c	856b	374**	237c	0.176	0.117	0.14
LSD		1%	5%				
(for comparison 2-V*S)	7.6		5.7				

a = Infection rate r) is the apparent rate of the daily increase in the percentage of diseased hills. In a column, means (average of four replications) followed by a common letter are not significantly different at the 5% level by Duncan's multiple range test.

** = significant at 1% level,

* = significant at 5% level,

ns = not significant

Table 3 : Area under disease progress curves (AUDPC) and apparent rate of tungro infection with different disposition of inoculum sources in IR22 in the 1993 Ds and WS (Satapathy *et al.*, (1996).

Source disposition (no. of foci/plot)	*AUDPC*		*Infection rate (r[a])*	
	1993 DS	*1993 WS*	*1993 DS*	*1993 WS*
0	11a	40a	0.88	0.051
1	47b	96ab	0.115	0.067
25	81c	125ab	0.125	0.068
100	94c	156b	0.119	0.067

[a]Infection rate (r) is the apparent rate of the daily increase in the percentage of diseased hills. In a column, means (average of four replications) followed by a common letter are not significantly different at the 5% level by Duncan's multiple range test.

Susceptible rice cultivars

Besides acting as good inoculum sources, susceptible rice cultivars are preferred by leafhopper vectors and help in their colonisation. Secondary spread of the disease becomes very fast in the susceptible cultivars and reaches epidemic level quickly. (Satapathy and Anjaneyulu, 1986; Yadav and Mishra, 1990). Resistance to tungro viruses is very rare in rice germplasm but resistance to the vector is quite common (Hibino *et al.*, 1990). These vector resistant varieties provide field resistance to tungro disease and it is due to non-preference and natibiosis associated with these vanities. Vector resistant varieties are likely to be less infected by *N. virescens* than susceptible ones. Moreover they support lower populations of vectors and these feed or relatively short periods. Feeding is mainly in the xylem rather than in the phloem which provides the main sites for virus acquisition and inoculation.

Furthermore, individual *N. virescens* on vector resistant varieties tend to be smaller and less fecund than those on susceptible ones. However these vector resistant varieties which were resistant at the time of release do not last long as their intensive and wide scale cultivation leads to the development of new virulent forms of the vectors that infect a previously resistant variety (dahal *et al.*, 1990a). IR22, IR36, IR54 are such examples. Varietal breakdown due to GLH biotypes has been observed in many countries such as Indonesia, the Philippines and Thailand and serious epidemics have been noted from time to time.

Young age of he rice plant

Rice plants at young age, irrespective of the variety are more susceptible to tungro viruses and their susceptibility comes down as plants get matured. (Rao and Anjaneyulu, 1980; Narayansamy; 1972). Besides enhancing disease severity and yield loss, early infected plants provide a greater amount of virus inoculum for disease spread than matured plants. Moreover young plants are preferred by green leafhoppers and a large number of eggs are oviposited thereby greatly contributing to the population build up of the vectors and disease spread (Lim *et al.*, 1974). Latent period (the time interval between infection and symptom expression) being short in young plants, it helps in increasing the number of infection cycles in a cropping season and thus contribute to the secondary spread of the disease (Rao and Anjaneyulu, 1979a).

Coincidence of susceptible rice cultivars virus inoculum, vector population, and young age of rice plant

Tungro disease appears in epidemic form only when virus inoculum coincides with young stage of susceptible rice cultivars and high vector population. The epidemic does not appear every year but at an interval of about 5 to 10 years when the above factors coincide. This is nature's balance. Had this appeared every year, probably complete devastation would have occurred.

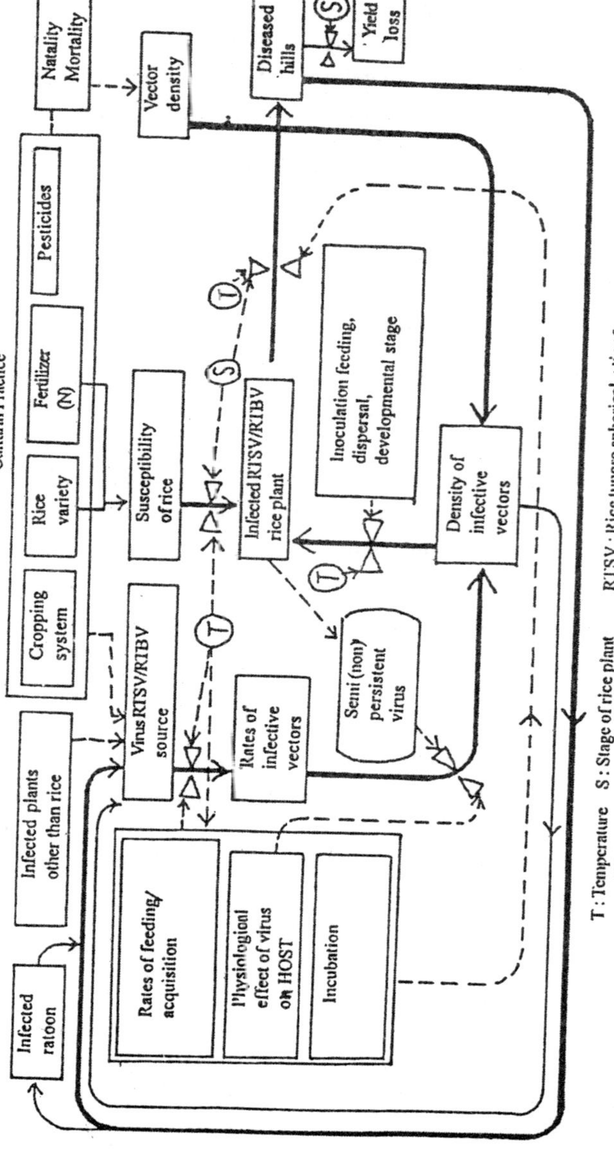

T : Temperature S : Stage of rice plant RTSV : Rice tungro spherical virus
RTBV; Rice Tungro Bacilliform virus.

Fig.14. Epidemiological cycle of Rice Tungro virus Disease

The leafhoppers appear every year on rice plants in tropical as well as temperate Asian countries. But virus inoculum becomes the limiting factor. When it becomes available along with susceptible varieties, tungro epidemic breaks out. However in tungro endemic areas virus inoculum is available almost round the year, but when susceptible cultivars become prevalent or earlier resistant variety breaks down, epidemic occurs bringing havoc and hardship to men and domestic animals.

Thus the stage of the plant growth when viruses are carried into rice fields is crucial in the development of tungro disease (Suzuki *et al.*, 1992). There is a greater potential for tungro disease spread in an early planted crop-provided that the amount of inoculum and the number of leafhoppers are not limiting. Moreover population size of leafhoppers in the plot is not important as disease spread is critically dependant on the number of infective leafhoppers entering into the plot (Savary *et al.*, 1993) or the amount of inoculum sources (Holt *et al.*, 1992; Chancellor *et al.*, 1996c) available. As a whole the susceptible nature of rice cultivars and the inoculum source are more important in disease spread than the other two components.

In simulation of tungro epiphytotic condition or experimental purpose, manipulation of above factors becomes essential. This has been done in the experimental form of Central Rice Research institute, Cuttack. This technique has helped the author (AA) to carry out several field experiments relating to epidemiology, disease management, insecticidal trial and varietal screening as explained earlier.

The simulation of tungro epidemic is carried at Cuttack during August-November coinciding with peak occurrence of green leafhoppers. Rice seeds are sown in August and the seedlings are transplanted in the first week of September about two months later than the surrounding fields. The inoculum is introduced about 5 to 7 days after transplanting and infection reaches at 100 per cent about 4 to 6 weeks after transplanting. The crop is harvested in December and the fields are kept dried and ploughed for the next season. This sort of simulation experiment can be carried out easily in places (states) where there is a single cropping season but difficult in places where hoppers remain active round the year and overlapping cropping practices are seen. The interaction of various factors in disease development has been shown in Fig. 14.

Cropping Pattern

As *N. virescens* is monophagous to rice and tungro virus sources are mostly confined to rice, cropping pattern highly influences epidemic outbreak. In asynchronously planting areas, in the absence of distinct cropping period and fallow period, inoculum is carried over and vector population remains high round the year thus helping in disease outbreak. In synchronous planting areas because of distinct planting time with fallow period and dry ploughing, risk of tungro outbreak becomes low (Holt and Chancellor, 1997). It helps to reduce vector buildup and virus inoculum sources in the volunteer seedlings and ratoon crops. As new biotypes of insect vector do arise due to intensive cultivation of a particular cultivar (Dahal *et al.*, 1990a) varietal diversification (growing many resistant varieties instead of one) and rotation (growing a set of varieties having different genes for resistance) has been found to reduce tungro disease (Manwan *et al.*, 1985; Habibuddin *et al.*, 1987). It reduces the selection pressure responsible for evolution of new insect biotypes. Keeping this in view, to prolong the durability of insect resistant varieties and to reduce the chances of new biotype development, a concept called gene rotation has been developed in South Sulawesi, Indonesia. In this scheme varieties have been classified into four groups, TO, T1, T2 and T3 based on their genetic background for GLH resistance. Varieties in To group are T1, Pacita etc which do not have GLH resistance genes. While T_1 (IR 26, IR30, IR46), T2 (IR32, IR36, IR38), T3 (IR29, IR36, IR50, IR54) groups have varieties with GLH1, GLH6 and GLH5 resistant genes respectively. Various groups of varieties are grown in the cropping season either singly or in various rotational schemes specific to localities to reduce disease incidence. Transplanted rice favours disease outbreak as compared to direct seeded rice Tiongco *et al.* (1990b) because the later favours natural enemy build up and thus reduction in vector population. Further in direct seeded rice, because of close spacing the viruliferous insects get limited opportunity for free movement and thus the chances of infection become low. Late planting both in wet and dry season promotes disease (Shukla and Anjaneyulu, 1981) It gives opportunity for the young and active tillering stage of rice crop to coincide with high vector population. Moreover infective (viruliferous) hoppers easily migrate from the old matured crop to young crops causing infection. Roguing does not reduce disease spread (Holt and Chancellor, 1996) as it disturbs the insects and their movement helps in further spread of the disease within the field.

Cultural Practices

Tungro disease spread is influenced by cultural practices besides nature of the variety, inoculum sources and amount of inoculation. Disease spreads very fast in susceptible varieties as compared to tolerant and resistant cultivars. In a varietal mixture, it is the proportion of susceptible varieties in the mixture of resistant and tolerant cultivars that influence disease development. Slow disease development occurs in close spacings especially in tolerant varieties due to reduced movement of insect vectors. Low disease occurrences in direct-seeded rice as compared to transplanted field has also been observed. (Anjaneyulu *et al.*, 1995).

Less disease incidence and slow infection rate has been observed under rainfed condition than under irrigated condition. Humidity and irrigation favours crop growth and population build up of green leafhoppers. Further irrigated fields attract more leafhoppers and favour movement of insect larvae causing further spread of the disease. Increase in disease incidence with increase in nitrogen level in the field has been reported by Shukla (1979). Subsequent studies on the effect have reported that effect of fertilizers, nutrients and soil amendments such ammonium chloride, ammonium sulphate, urea, diammmium phosphate, zinc sulphate had no effect on maximum disease evidence though they effect growth and yield to certain extent (Anjaneyulu *et al.*, 1995).

Insecticide application has been found effective in reducing tungro disease spread (Satapathy and Anjaneyulu, 1984). Insecticides kill the insect vectors quickly and prevent virus transmission. However use of ineffective insecticides or sub-lethal doses irritates insects causing further spread of the disease instead of its control.

Rice Ecosystems vs. Tungro Epidemics

Rice is a semi-aquatic cereal which originated in the tropics where vast areas of flat, low-lying land are flooded annually during the monsoon season. Rice is perhaps the only major food crop that can grow with its roots under water. The centre of rice cultivation remains the lowland humid tropics, but owing to human selection and natural dispersal, rice is now cultivated as far north as the banks of the Amur River (53°N) on the border between the USSR and China, and as far south as Central Argentina (40°S). Rice is also grown in cool climates in the mountains of Nepal and India, and under irrigation in the hot deserts of Pakistan, Iran and Egypt. It is an upland crop in parts of Asia, Africa, and Latin America. At the other environmental extreme are floating rices, which thrive in the seasonal deep flooding of large river deltas the Mekong in Vietnam, the Chao in Thailand, the Irrawaddy in Burma, and the Ganges-Brahmaputra in Bangladesh and eastern India. Rice performs better than other grain crops in areas with saline, alkali or acid sulphate sails. Clearly rice adapts well to diverse growing conditions. It is now harvested from more than 150 million hectares, over 10 per cent of the world's arable land. Total production is about 420 million tons of unhusked rice yielding about 275 million tons of milled grain.

Because rice is grown under diverse ecological conditions, cultural and production practices vary considerably and adopted varieties have distinct characteristics. Rice growing environments have been classified into five major categories, irrigated, upland, rainfed low land, deepwater and tidal wet lands. These major categories have been divided (IRRI, 1985) into several subcategories as shown in Table 4.

Table 4 : Classification of rice growing environments (area of each major capacity indicated in parentheses) (IRRI, 1985)

Irrigated (77 million ha) Ecosystem

Irrigated, with favourable temperature

Irrigated, low temperature, tropical zone

Rainfed lowland (33 million ha) Ecosystem

Rainfed shallow, favourable

Rainfed shallow, drought-prone

Rainfed shallow, drought and submergence-prone

Rainfed shallow, submergence-prone

Rainfed medium-deep, waterlogged

Deepwater (12 million ha) Ecosystem

Deepwater (50 cm to 100 cm water depth)

Very deepwater (100 cm water depth)

Upland (19 million ha) Ecosystem

Upland, long growing season, favourable soil factors

Upland, long growing season, unfavourable soil factors

Upland, short growing season, favourable soil factors

Upland, short growing season, unfavourable soil factors

Tidal wetlands (5 million ha) Ecosystem

Tidal wetlands with perennially fresh water

Tidal wetlands with seasonally or perennially saline water

Tidal wetlands with acid sulfate soils

Tidal wetlands with peat soils

In irrigated rice areas, water level is regulated to provide optimum water supply to rice crop. The water is supplied either through canals, from deep wells or tanks depending on its availability and need.

In rainfed rice, the crop depends directly on rain water and there is no regulation of water level. The crop grows in uplands with plots bunded to store rain water. In deep water areas there is stagnation due to poor drainage system and rice can grow in places having water level upto 30ft.

Rice tungro can occur in epidemic form in any condition starting from rainfed uplands to deep water rice. Since tungro viruses are transmitted by green leafhopper *N. virescens*, its preference for as well as population size in a particular ecosystem determines the level of infection.

Rainfed Upland and Irrigated Ecosystem

About 19 million hectares are planted to upland rice which is about 15 per cent of the world's rice area. In tropical Asia alone, more than 11.5 million hectares of varying topohydrologica-edaphic regimes are planted to upland rice. Under this ecosystem, rice is direct seeded in non-flooded, well drained soil on level to steeply sloping fields. Crops suffer from lack of moisture and adequate nutrition and yields are very low.

In irrigated ecosystem rice is transplanted or direct seeded in puddled soil on levelled, bunded fields with water control, in both dry and wet seasons in the lowlands, in the summer in higher elevations, and during the dry season in flood prove areas. The crop is heavily fertilized and yields can reach 5 tons per hectare in the wet season, more than 8 tons in the dry season. Irrigated rice is planted on 81 million hectares worldwide, accounts for 55 per cent of the world's harvested rice area and contributes 76 per cent of global rice production. Irrigated systems in Asia are concentrated in the semiarid and subhumid subtropics.

In both these ecosystems, rice crop is subjected to numerous pests and diseases and tungro is one of them.

Spread of tungro disease in rainfed upland and irrigated areas has been studied on nine high yielding rice cultivars, TN1, IR20, IR26, IR30, Jaya, Ratna and CR 138-994-27 by Shukla and Anjaneyulu (1982b). The disease incidence and rate of infection were remarkably less under rainfed than irrigated condition in all nine cultivars. The slow rate of disease spread under rainfed condition was attributed to low vector population. High disease incidence in wet nursery bed condition than dry beds has also been observed (Shukla, 1979).

Generally, irrigated ecosystem is the most ideal for fast spread of tungro disease to reach epidemic level. The humidity and optimum temperature are very conducive for vector multiplication. Actively growing rice plants and their leafsheaths provide ideal environment for the females to lay eggs. Egg hatching becomes quicker. Moreover the water level provides an ideal surface for jumping of nymphs and spread of the disease among luxuriously growing plants. In these ecosystem, to maintain yield stability, a lot of stress is being given for incorporating genes for disease and insect resistance.

Rainfed Lowland Ecosystem

Rainfed low lands (rainfed plants with shallow water). Rice is transplanted or direct seeded in pudded soil on level to slightly sloping, bunded or diked fields with variable depth and duration of flooding, depending on rainfall. Soils alternate from flooded to non flooded. Yields very depending on rainfall, cultivation practices and use of fertilizer. Rainfed lowland rice is grown on 37 million hectares worldwide. It is dominant in the humid and subhumid tropics.

It is more or less comparable to irrigated ecosystems and provide very suitable environment for multiplication of leafhopper vectors and tungro disease spread. Rainfed low land ecosystem occupies a dominant area in India (eastern states like Orissa, Bihar, and West Bengal), Bangladesh and parts of Thailand. If tungro appears in epidemic from in these areas, it can prove disastrous.

Deepwater Ecosystem

Excess water is as important as insufficient water in limiting the adoption of improved rice varieties. Rice is direct seeded or transplanted in the rainy season on fields characterized by medium to very deep flooding (50 to more than 300 cm) from rivers and from tides in river mouth deltas. Soils cycle from dry to flooded and may have severe problems of salinity and toxicity. The crop grows as flood waters rise, with harvest after the floods recede. Flood prone rice grown on more than 10 million hectares predominantly in south and southeast Asia.

In this ecosystem rice plants grow with major part of their stem/leaf submerged in the water. This is not a good environment for disease spread as GLH fail to multiply due to lack of sufficient leaf sheath area for them to lay eggs. In these areas, the primary infection by adult leafhoppers predominates. Secondary infection by nymphs is very much limited. During 1990 epidemic in eastern Indian states the author (AA) has observed vast stretches of water bodies (with about 1m level of water) with rice crop having only few leaves visible above. All the leaves were orange coloured and on transmission test found to be tungro infected. The question comes whether the vast strethes of water were more attractive for the leafhoppers to settle for rest and to feed on. This is a subject for investigation. It is interesting to note that there was 100 per cent infection in this ecosystem. It is known that deep water ecosystem contributes a very little to the total rice yield due to unfavourable agroclimatic conditions. This poor yield condition can further be aggravated if tungro disease occurs very frequently in these large areas.

Tidal and Wetland Ecosystem

In this ecosystem there is about 5 million ha of rice crops in tidal wetlands, three types of floods are superimposed on each other in ratios that depend on the relative distance of the site from the sources of flooding: (i) diurmal flooding resulting in mixing of salt water with fresh water; (ii) fortnightly tides from the interaction of fresh water supplied by rivers with moon tides from the sea; and (iii) monsoon-derived fresh water floods from rivers. In tidal wetland ecosystem, tungro disease is rarely seen as saline environment does not favour for insect vector multiplication.

Climate, Vector and Virus

Plant disease epidemics develop as a result of the timely combination of some elements such as susceptible`host plant, a virulent pathogen and favourable environmental conditions over a fairly long period of time. Further human activities sometimes initiate, develop or stop epidemics.

The epidemiology of tungro disease is very interesting because of the presence of insect vector (*Nephotettix* sp.) besides the three traditional components host, pathogen and environment of the disease triangle. Presence of different rice cultivars, vector biotypes, virus strains, different secondary hosts, rice ecosystems, and their interaction make tungro epidemiology very complicated.

Climatic factor (environment) in terms of rainfall, humidity, maximum and minimum temperature, sunshine hours, soil nutrition, wind speed etc. affect the host as well as the vector which in turn influence the virus spread and outbreak of epidemics.

The population dynamics along with annual fluctuation of green leafhopper population has been discussed earlier. From Cuttack the peak period of leafhopper vector (*N. virescens*) during September-October has been noted to be associated with moderate range of maximum-temperature (31.5-32.7°C), and high range of minimum temperature (21.9 to 25.3°), high relative humidity

(74-85%) and periods followed by heavy rainfall. The low density of leafhopper population during dry season (May-June) is attributed to high temperature and wide range of relative humidity. Lowe and Nandi (1972) have observed that both heavy and low rainfall are not favourable for vector buildup whereas light and constant rainfall and periods following extremely heavy rainfall helps in rapid build up of vectors. Moreover a variable lag period has been observed between rainfall and appearance of leafhoppers (Mukhopadhyay and Mukhopadhyay, (1986). Changes in light illumination of the lunar cycle has been noted to affect GLH activity. The population of green leafhopper species rises in the first tour phases after the new moon, followed by a decline and then again an increase between first quarter and full moon, the increase being more marked in *N. nigroopictus* and *N. virescens.* Mukhopadhyay (1986) has observed that amount and distribution of moon and *N. viresans* light during lunar cycle provides photo period cues and. synchronization of light activity has important consequences in the pattern of crop infestation and spread of tungro disease.

Climatic factors strongly influence leafhopper biology. It has been observed from Cuttack (India) that maximum (31.8°C) and minimum (21.3°C) temperatures, high relative humidity (83.5%) followed by high rainfall, continuous low rainfall and low sunshine hours (5-2 hr) are favourable for oviposition and egg hatching. Oviposition usually occurs during the day time. In ideal temperature (20°C) the incubation period is about 7-9 days and the eggs hatch mostly between 6.00 to 19.00 hours. The average total nymphal period is about 16 to 19 days (Siwi *et al.*, 1987) each instar lasting about 3 to 4 days. The average life span of an adult insect is about 25-26 days (Cheng and Pathak, 1971) but sometimes it reaches upto 46 days (Alam and Islam, 1959). The females survive longer than the males. The developmental stages and adult longivity are influenced by temperature and a temperature of 29-33°C is optimum for GLH. The period required for completion of generations during June-September is much less than that required for other periods of the year thus wet season favouring vector buildup and higher disease incidence (Chakravarti *et al.*, 1979). In warm and humid tropics, *Nephotettix* sps remains active round the year and their population fluctuates according to the available host plants, environmental conditions, natural enemies and cropping pattern. When rice crop approaches maturity, insects migrate to young crop or grasses. However, in areas of wide temperature variation, they hibernate or aestivate. The hibernating insects become active when weather warms up around March to April and migrate to grasses where they breed for one generation before migrating to rice fields shortly after transplanting. In areas where rice is available at the time of hibernation, the insects migrate directly do rice fields. Thus the seasonal occurrence varies distinctly between areas where the insect undergoes dormancy and diapause on the one hand and where they remain active throughout the year on the other hand. Disease spread has also been found to be influenced by vector movement which in turn is influenced by temperature (Miah and Long, 1978), velocity and direction of wind (Rahman *et al.*, 1985).

Agroclimatic factors influence cropping pattern as well as the host (rice) plant that act as virus source. It has been observed that more plants become viruliferous and more infection occurs during aman season as compared to aus and boro (Mukhopadhyay and Chowdhury, 1973). With rise in temperature efficiency of transmission and acquisition increases. Further symptom expression and its severity are also said to be influenced by seasons. Plants exhibit orange discolouration during monsoon (July-October) whereas in summer and winter they appear dull orange and orange respectively (Rao and Anjaneyulu, 1982). The cultivar Jaya exhibits severe tungro symptoms in the early days after infection and the plants slowly recover and appear green subsequently during summer and monsoon, but in winter the infected leaves exhibit chlorosis. Greater disease incidence is usually observed ruing the wet season than in the dry season in different rice growing areas because in the wet season leafhopper population remains high and more areas come under rice cultivation. Tungro disease incidence usually reaches high level during the peak period of vector population in the year and is associated with intermediate planting dates. Early and late planted rice fields escape infection as in the former, the plants get physiologically matured and become resistant to virus infection whereas in the later, the peak period of leafhopper is reached before the crop is planted. From monthly plantings at Cuttack it has been noted (Shukla and Anianeyulu, 1981a) that tungro incidence is high is September, October, November and December plantings, while disease was either negligible or absent in other months.

Tungro Endemic and Tungro Epidemic Prone Areas

Tungro disease is a problem mostly in tropical islands such as Indonesia, Malaysia, the Philippines, Thailand and in coastal areas of India and Bangaladesh. Green leafhopper vector is one of the prime factors responsible for tungro epidemic. The vector multiple in the environment where there is an optimum temperature throughout the year coupled with high humidity.

Looking into the geography—the equator passing through Indonesia runs very close to Malaysia. In these islands, the temperature almost remains the same round the year. The Philippines, Thailand and southern India are also close to the equator. In these regions rice is grown throughout the year and vector also becomes available. Consequently some of these locations manifest as endemic to tungro.

Another factor behind endemism is overlapping of rice crops. Rice appears to be the prime host, both for viruses and vector *N. virescens* (though it may not be absolutely monophagous). There is not a single weed species which is known to act as an alternate host, though there are some reports about weed susceptibility under experimental conditions. Very few weeds have been reported to be natural hosts of the viruses but none of them have shown clear symptoms and transmission of viruses have failed. Consequently tungro endemic areas are localised where there is overlapping of rice crops due to intensive cropping. Overlapping of crops lead to the transmission of viruses from diseased to healthy plants. Asynchronously planting helps in the transfer of inoculum and its perpetuation, sometimes stubbles of infected plants, ratoons, volunteer rice plants favour multiplication of green leafhopper vectors and the viruses. It leads to recurring of tungro disease. The exact nature of tungro endemism needs further study and analysis.

In West Bengal, an endemic state for tungro in India. Overlapping of rice is seen with three cropping seasons, aman (July-September) boro (Nov-March), and aus (April-July). Intensive cropping has resulted in the availability of virus inoculum throughout the year. Mukhopadhyay (1984a) observed that nursery beds raised just before or after the harvest of rice crop play a key role in perpetuation of the virus in overlapping cropping areas. Tungro epidemics are been often noted from this state. Eversince tungro is identified in India in 1968, severe epidemics have been reported in 1969, 1973, 1984 and 1990. Besides these severe epidemic years, moderate level epidemics have occurred in between with epidemics of minor nature occurring almost every year in some locations or the other. Besides West Bengal, endemic pockets are also seen in certain locations of Assam and Tripura in India and in Bangladesh (country close to West Bengal in India).

Prior to 1968, there might have been severe epidemics. For example there was a great Bengal famine in 1942 with a lot of starvation death. At that time, this famine was attributed to *Helminthosporium* a fungal rice disease because knowledge on tungro was total lacking then. Usually *Helminthosporium is* a weak pathogen and generally occurs secondarily in nutritionally depleted rice leaves. Because *helminthosporium* often occurs in tungro infected plants in experimental plots, it is interpreted that in 1942 great Bengal famine primarily might have been due to rice tungro disease and secondarily due to *Helminthosporium*. Moreover beyond 1942 *Helminthosporium* has never been reported as a disease appearing in epidemic form though West Bengal has experienced tungro epidemics at an interval of every 3-4 years repeatedly.

Besides West Bengal there are certain pockets in Andhra Pradesh (Nellore district) and Tamil Nadu where tungro endemism is also observed.

Tungro epidemic prone areas are those where tungro epidemic occasionally not at a regular interval. States of Orissa and Bihar are the examples of this category. Major epidemics had been recorded in these states in the years, 1969, 1981 and 1990. Green leafhoppers are seen only during monsoon season in these states but disappear with the approach of winter due to cold. Consequently chances of being endemic becomes low.

The sporadic appearance of the disease is due to sudden appearance of the inoculum in the rainy season when maximum areas are brought under rice cultivation and leafhopper vectors appear in the nature in huge number. The source of inoculum may be some unknown weeds growing as alternate hosts in the region or infective (viruliferous) leafhoppers coming though distant migration by the help of winds, cyclone or lights of running vehicles.

In endemic areas, varieties often become susceptible due to appearance of new vector biotypes and it needs frequent charge of varieties of course varietal rotation makes them durable.

In epidemic areas due to sporadtic appearance of the disease farmers do not show much concern for the disease and do not like so change the varieties. Consequently sudden appearance causes huge loss bringing sometimes famine condition. The epidemic and endemic pockets of tungro needs further investigation with emphasis on vector migration and existence of alternate host(s) if any.

Risk Factors in Tungro Epidemics

Tungro disease strikes suddenly and vanishes subsequently affecting thousands hectares of rice crop. A number of factors are said to be linked with risk of tungro (Hot *et al.*, 1996). These factors include nature of rice variety(s) cultivated, planting time, cropping frequency, inoclum source and vector abundance. Some of these factors are field specific whereas others operate at a wide.

It has been noted that higher tungro incidence occurs in wet season than dry, and disease becomes severe in susceptible varieties as compared to cector resistant because of low virus transmission in the later. The risk of tungro incidence is the highest in late plantings both in wet and dry seasons (Shukla and Anjaneyulu, 1981; Chancellor *et al.*, 1996b). This is due to that late season plantings are exposed at a suseptible growth stage to increasing amoutns of both inoculum and vectors which had built up on earlier plantings, consequently putting the late plantings in the greatest risk of tungro incidence When seedlings are planted early, the plants get physiologically matured and escape infection by the time the vectors appear in the nature. As far as possible the farmers may be advised to go for early plantings to prevent tungro disease and so as to reduce the carry over of inoculum and leafhopper vectors between seasons as well.

Further intensive cultivation of one or few varieties continuously lead to the appearance of new vector biotypes that infect a variety that behaves as resistant earlier. Consequently the varietal resistance breaks down. Failure of varietal resistance has led to major epidemics in the past in the countries like philippines. Under such situations, farmers may be advised to grow a couple of vector resistant varieties in rotation instead of one to promtoe durable resistance and to help escape crops from epidemics.

Besides vector abundance, the proportion of infective vectors (infection sources) play very important role in disease spread in endemic areas as the disease spread is associated with vector numbers (Suzuki *et al.*, 1992; Savary *et al.*, 1993). In non-endemic areas, it is the proprotion of infective vectors in the population play very critical role in spreading the disease. Thus vector size do not have important role once tungro infection is established in an area. Vectors are a necessary but not sifficient condition for disease spread. Of the vector and inoculum sources, the latter plays the import role as insects appear all the years but when inoculum becomes available and it coincides with active growth stage of susceptible varieties, disease spreads very fast. Control of the inoculum sources/prevention of infection is more important than disease control.

Asynchronous planting favours carry over of inoculum and insect vectors and increases the risk of tungro epidemic whereas synchronous planting with a fallow (crop free) period decreases it. A crop free period with dry ploughing not only promotes absence of rice crop but also prevents the growth of stubbles, ratoons and volunteer rice plants that harbours insect vectors and inoculum. Synchronous planting for some years lowers the inoculum pressure and reduces the risk of tungro epidemics. Moreover a field located within 1 km distance from an inoculum source has a high risk of tungro and the risk comes down as the distance increases beyond

1 km (Holt *et al.*, 1996a), may be due to short range flight of *N. virescens.* In epidemics affecting large areas, the infective vectors probably get widely dispersed initiating the disease in several fields in a locality from where the disease spreads secondarily to nearby areas stepwise through short flights again. Long distance migration of few infective hoppers or sometimes winds favouring dispersal of vectors might be helping in the breakdown of epidemics.

13

Disease Forecasting

Besides causing serious yield losses, major problem associated with tungro is its highly unpredictable nature. It strikes suddenly and vanishes thereafter for couple of years. Consequently forecasting of the disease is highly essential to develop effective management strategies.

Studies from India has shown that there is no correlation between meteorological factors and tungro spread (Vidyasekaran, 1985), Suzuki et al (1993) observes that rainfall affects tungro virus transmission and the relationship is very complex. Vidyasekaran and Lewin (1986) have reported that in epidemic years vector population increases into very high compared to non-epidemic years. Many others (Lim 1972; Hino *et al.*, 1974; Shukla and Anjaneyulu, 2982a) have noted a direct correlation between vector population and disease incidence. However Chancellor *et al.* (1996c) observed that vectors are a necessary but not always a sufficient condition for disease spread. While low inoculum availability limits disease spread, high vector numbers may or may not be associated with a high tungro incidence. From analysed survey data collected from three sites in the Philippines, Savary *et al.* (1993) observed that in nonendemic areas tungro outbreak were mostly associated with occurrence of infective vectors (virus inoculum), and the size of leafhopper population had little impact whereas in endemic area tungo incidence is associated with large number of infective vectors as well as high leafhopper population. Suzuki *et al.* (1992) reported that tungro outbreaks are triggered by sporadic occurrence of severely infected paddy fields in RTVD-endemic areas around the beginning of the wet season. Subsequently they (Suzuki *et al.*, 1997) found that tungro incidence in a newly infected field is due to new infections in the previous month, occurrence of crops 2-3 months old and rainfall in the current month. Hence assessment of inoculum source as well as vector population is essential to predict tungro outbreak. In Malaysia and some other countries, during cropping seasons surveillance activities through the use of light traps and mobile nurseries are intensified to monitor vector population to predict disease outbreak. While light traps give information about leafhopper population, mobile nurseries (trays with young seedlings) placed in fields prior to planting help to detect the presence of viruliferous insects through ELISA indexing of seedling. Areas are also constantly monitored to detect the presence of infected hills. As and when disease outbreak is predicted, extension workers are alerted through an expert system to decide on the appropriate control measures to be undertaken. Risk factors once known help to decide and take appropriate preventive measures to minimise disease spread.

Strategic Management

Tungro disease appears suddenly, spreads very fast covering large areas causing alarming situation and vanishes thereafter. Usually the disease recurs at an interval of 5 to 10 years but the exact time of its appearance is difficult to predict. In the absence of correct forecasting system the disease management has become a difficult task. The main strategy for control of tungro has been cultivation of vector-resistant varieties but many such varieties which were field resistant at the time of release have succumbed to the disease as vector population become adapted to them in course of time (Dahal *ét al.*, 1990a). Vector control through insecticides has commonly been recommended (Satapathy and Anjaneyulu, 1986, 1989a,b) but it has helped in the development of insecticide resistant green leafhopper vectors besides pest resurgence, killing of non-target organisims and environmental pollution (Litsinger, 1989; Chancellor *et al.*, 1997; Statapathy, 1998). Some cultural practices such as plant spacing, roguing, conservation of natural enemies manipulation of planting time etc. have often been recommended for controlling disease but some of them are difficult to implement and others do not provide dramatic results (Heinrichs, 1979). In this background appropriate management strategy is to be followed for tungro disease control keeping the risk factors in view.

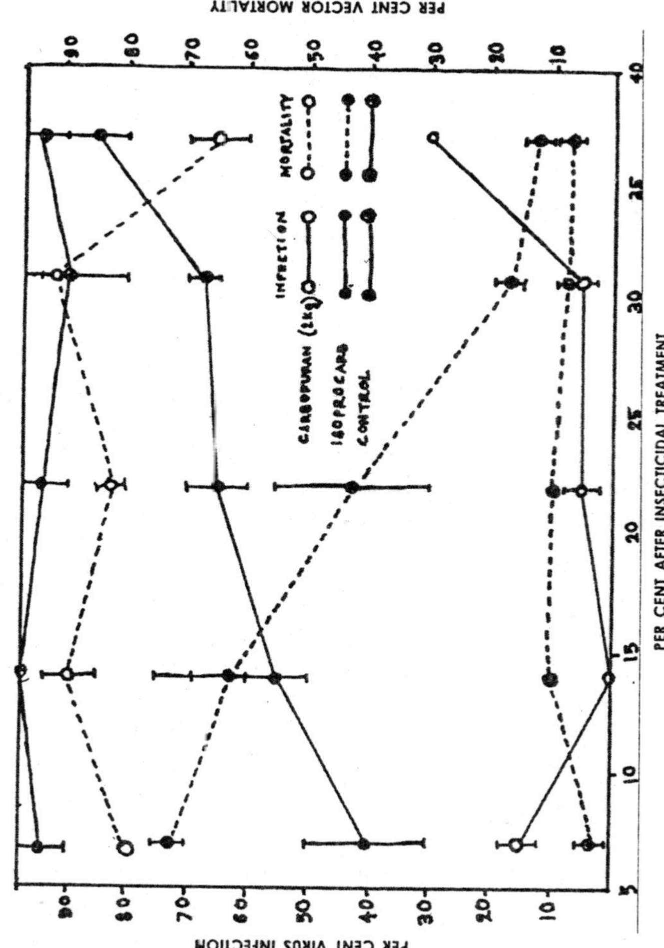

Fig.15. Persistency of granular insecticides against virus infection and vector mortality by soil incorporation method.

In endemic areas, though tungro appears every year, the areas infected as well as disease severity vary from year to year. In these areas it is necessary to grow tungro resistant/tolerant cultivars preferably RTSV resistant varieties. These RTSV resistant varieties can be identified either through transmission, or serological test or both. Consistent cultivation of RTSV resistant or RTBV tolerant cultivars not only control tungro disease by reducing its secondary spread but will help in reducing tungro inoculum so that the epidemic can be restricted in the future.

Many varieties released in the past for tungro are mostly resistance to the insect vector but not to the virus(s). The GLH resistance is said to be associated with non-preference (reduction in feeding) or with antibiosis (reduction in reproduction, development etc) or both, while non-preference reduces the influx of viruliferous vectors, antibiosis restricts secondary spread through reduction in vector abundance. Because insects do little feeding on varieties with high levels of resistance, the crop escapes infection. But these varieties when cultivated continuously for years break down in their resistance due to development new GLH biotypes that now feed on these verities. These varietal break down in resistance has been reported from Thailand. Indonesia and the Philippines. Varieties such as IR36, IR42, IR50, IR54, IR64 which were almost free from infection at the time of their release in the Philippines do not stand resistant to tungro in the field now (Dahal *et al.*, 1990a).

Identification of RTSV resistant variety is easy. A good number of varieties resistant to RTSV have been detected at IRRI, (Tables 5 and 6) Philippines (Hibino *et al.*, 1990; Koganezawa and Cabunagan, 1997). Based on 25 years of field experience on varietal screening the author (AA) believes that about 20 per cent of the germplasm and 10-15 per cent of high yielding varieties and breeding cultures are resistant to RTSV. They do not exhibit clearcut symptoms except mild stunting. Due to regular epidemics, there is also natural selection of RTSV resistance in endemic areas. Farmers can also easily identify tungro (RTSV) resistant cultivars from other varieties grown in their locality through cultivar reaction. For example in 1990 epidemic, in coastal districts of Andhra Pradesh, MTU 5923 withstood the epidemic and subsequently found to be RTSV resistant through transmission test. Consideration should be given to find out the most effective way of deploying resistant varieties. Since cultivation of specific RTSV resistant cultivars is like to generate new strains of the virus (Anjaneyulu, *et al.*, 1997) vector resistance is being combined with RTSV resistance/or RTBV tolerance to prolong their useful life. However early planting along with rotation of groups of RTSV\varieties looking to farmers adaptability is likely to enhance their durability. However limited availability of specific RTSV resistant seeds or their high cost or farmer's limited access to such varieties may stand on its adoption. Moreover some farmers may continue to grow susceptible varieties because of their superior eating quality or high yield potential. Under such circumstances, use of systemic granular insecticides such as carbofuran may help in reducing tungro disease incidence (Satapathy and Anjaneyulu, 1990) through vector control.

Table 5 : Rice varieties showing resistance to infection with rice tungro spherical virus (RTSV) (Koganezawa *et al.*, 1997).

Accession number	*Variety*	*Accession number*	*Variety*
177	Adday Sel. *	26789	Shalya+
180	Adday Local Sel.*	26791	Sham Rosh*
4021	Binicol+	26813	Gogoi*
5999	Pankhari 203+	27529	Bhoilush+
7366	PL 184675-2+	27779	Bara Pashwari 390
8261	Padi Kasalle	27781	Bara 143+
11062	G 378	27787	Basmati Nahan 381
11751	Habiganj DW 8	27798	Basmati 1
12203	ARC 6064+	27799	Basmati 43 A+

(contd...)

1	*2*	*3*	*4*
12274	ARC 6561	27800	Basmati 93+
12310	ARC 7007+	27803	Basmati 107+
12428	ARC 10312	27804	Basmati 113+
12437	ARC 10343	27805	Basmati 122+
14504	IR 580 420-1-1-2	27814	Basmati 208+
14527	Barah	27818	Basmati 242
14649	Grendjah Melati	27821	Basmati 370 A+
14703	CPA 86805-2+	27828	Basmati 376+
15769	Lawangeen+	27829	Basmati 377+
16680	Utri Merah*	27830	Basmati 388
16684	Utri Rajapan	27832	Basmati 405
19680	ARC 10963	27833	Basmati 406
20600	ARC 7321	27835	Basmati 427+
21164	ARC 10980	27836	Basmati 433
21310	ARC 11315	27856	Basmati 302
21337	ARC 11346	27869	Chahora 144
21342	ARC 11353	27870	Chahora 148
21473	ARC 11554+	27872	Chahora 292
21474	ARC 11555	27873	Chahora 382
21745	ARC 11920*	27916	Dhanlu 254
21958	ARC 12170	27943	Hansraj 54+
22176	ARC 12596	27946	Hansraj 62
22199	ARC 12620	27947	Hansraj 189
22215	ARC 12636	27948	Hansraj 197
22307	ARC 12746	27951	Hansraj 365A
22331	ARC 12778	28102	P 590

(contd...)

1	2	3	4
26253	Nep Bap*	28302	Toga 286 A+
26295	Bale Betor+	28341	9+
26316	Birpla+	28450	361 +
26410	Pala Bhir*	28522	Gundrikbhog
26418	Shada Muta+	28867	AUS 4
26495	Konek Chul+	31746	Bish Katari+
26527	Shuli 2	36731	Firro E(1)
26560	Baharat +	37215	Matichakama
26582	Buchi 2	37337	Urman Sardar
26622	Gia Dhan+	37430	Ghigos
26633	Gurdoi+	37482	Kanakchul
26663	Kaisha Binni+	37488	Kashiabinni+
26703	Kurki+	37491	Katijan+
26715	Lao Bhug+	37761	Maliabhangor 1096*
26784	Sakort+	49996	Ovarkondoh*

* Cultivars showing resistance to both RTSV strains A and Vt6

\+ Apparent resistance to RTSV may be due to vector resistance.

Table 6 : Rice varieties showing tolerance to infection with rice tungro bacilliform virus (RTBV) Koganezawa *et al.*, 1997).

Accession number	*Variety*	*Accession number*	*Variety*
177	Adday Sel.	21956	ARC 12168
180	Adday Local Sel.	21958	ARC 12170
3707	Andi fr N. Pokhara	22176	ARC 12596
4021	Binicol	22199	ARC 12620
5346	Seratus Hari T 36	22215	ARC 12636

(contd...)

1	2	3	4
7366	PI 184675-2	22309	ARC 12748
11751	Habiganj DW 8	22331	ARC 12778
12203	ARC 6080	26418	Shada Muta
12207	ARC 6561	26468	Hansa
12274	ARC 7007	26494	Kola Mona
12310	ARC 10321	26495	Konek Chul
12428	ARC 10343	26527	Shuli 2
12437	Barah	26582	Buchi 2
14527	Lawangeen	26622	Gia Dhan
15769	Tjempo Kijik	26663	Kaisha Ninni
16602	Utri Merah	26682	Kashna Binni
16680	Utri Rajapan	26703	Kurki
16684	Balimau Putih	26772	Pura Binni
17204	Betrik	26776	Raja Mun
17292	ARC 5905	26791	Sham rosh
19675	ARC 10963	22831	Gogoj
19680	ARC 7140	27821	Basmati 370A
20533	ARC 7321	27827	Basmati 375A
20600	ARC 11346	27828	Basmati 376
21337	ARC 11353	31746	Bish Katari
21342	ARC 11355	37430	Ghigos
21344	ARC 11554	37473	Kalakura
21473	ARC 11555	37488	Kashia Binni
21474	ARC 11558	37491	Katijan
21476	ARC 11698	37509	Kushiyari
21569	ARC 11920	49996	Ovarkondoh

In tungro prone (non-endemic) areas, cultivation of resistant/ tolerant varieties is desirable. A number of varieties tolerant to RTBV has been given in Table 5. But farmers may not like to grow resistant varieties because of uncertainty of the disease outbreak and quality of the variety and their own choice. Under such conditions for moderately resistant or susceptible varieties where there is a clearly identifiable risk of tungro disease based on seasonal factors and disease incidence record in the locality, use of suitable insecticides may be a viable control option for farmers (Chancellor *et al.*, 1997). Besides suggesting for early planting, the effective control of tungro spread through insecticides is timely and appropriate application of fast acting quick knockdown and long persistence insecticides A list of recommended insecticides along with their time of application has been given in Table 7. Carbofuran is a promising insecticide found effective against tungro in the field in many countries (Shukla and Anjaneyulu 1980; Mochida *et al.*, 1986; Habibuddin *et al.*, 1987, Satapathy and Anjaneyulu 1989a). Because of its long residual toxicity (Fig.15) and fast action has been found promising in different methods of application such as broadcast, root zone and soil incorporation methods. Soil incorporation method has been found to be very effective (Table 8) in reducing infection and increasing yield besides reducing waste and risks associated with rainfall, dilution or flowing away in paddy water and removal from the foliage.

Cypermethin a synthetic pyrethoid has been found very effective both under transplanted field and nursery bed conditions (Satapathy and Anjaneyulu, 1982c). It has been effective in reducing tungro incidence to 3.0 and 0.3 per cent in susceptible cultivar TN1 and moderately resistant Ratna (Table 9). No adults and nymphs are seen in cypermethrin treated plots.

Though FMC 35001, phosphormidon and demeton-o-methyl sulfoxide are good in tungro disease control, they are less effective than cypermethrin. Cypermethrin at 0.005 per cent concentration was superior to broadcast application and almost equal to root zone treatment of carbofuran in disease control (Satapathy and Anjaneyulu, 1990). While selecting and using insecticides issues such as factors relating to human health, emergence of vector resistance, resurgence of secondary pests, environmental pollution etc. should be considered (Satapathy, 1998) as far as possible.

Table 7 : Schedule of insecticidal application when tungro symptoms are first seen in the fields or in the neighbouring fields, coupled with 2-3 or more leafhopper per plant, at different growth stages of rice crop.

Crop state (DAT)		*Insecticide*
Nursery	:	Carbofuran (3 kg a.i/ha) (20) and Cypermethrin
0-15	:	Carbofuran (15), Cypermethrin (7), Monocrotophos (10) and Cypermethrin.
15-30	:	Carbofuran (15), Cypermethrin (7) and Monocrotophos
30-45	:	Carbofuran (10-15) and Cypermethrin
45-60	:	Cypermethrin
60 and beyond	:	No spray Age induced resistance takes care.

DAT : Days after transplanting

Dosage : Carbofuran 1kg a.i/ha, Monocrotophos 0.05% and Cypermethrin 0.01%.

Figure within the bracket after insecticides reveals the number of days, the farmer has to wait until the next spray.

Note:
1. If Monocrotophos is not available, any systemic insecticide may be used.
2. If Cypermethrin is not available, any synthetic pyrethroid may be used.
3. Since continuous application of synthetic pyrethroid may cause resurgence, one spray of systemic insecticide is recommended between two sprays of synthetic pyrethroid.

Table 8 : Disease incidence, grain yield and leafhopper population in Taichung Native 1 (T) and Jaya (J) rice treated with carbofuran by different methods (Satapathy and Anjaneyulu, 1991)

Mode of application	*Dose*	*Total a.i./ha (kg)*	*Disease incidence (%)*		*Grain yield (kg/ha)*		*Leafhoppers (No,/20 hills) a*			
							Adults		*Nymphs*	
			T	*J*	*T*	*J*	T	J	T	J
S.I	2.0 kg a.i./ha	2.0	0.8 ab	0.8 ab	4724 b	5964 bc	0.0a	0.7a	7.3a	8.7b
A.1	4.0 kg a.i./ha	4.0	0.6 ab	0.3 a	5813 a	6870 a	0.0a	0.0a	8.0a	8.0b
R.Z	0.5 kg a.i./ha	0.5	2.5 b	2.5bc	4308 bc	5413 cde	1.3 a	2.0ab	12.7a	4.0ab
R.Z	1.0 kg a.i./ha	1.0	2.0 ab	2.4	4604 b	5891 bcd	0.7a	2.7ab	6.0a	2.7ab
R.Z	2.0 kg a.i./ha	2.0	1.2 ab	1.4 ab	5829 a	6312 ab	0.0a	0.0a	2.7a	0.7a
F.S 5 times	0.1%	5.0	18.3 d	13.7 d	3412 d	4707 e	0.0a	0.7a	2.7a	0.7a
B.C. Thrice	2.0 kg a.i./ha	6.0	12.3 c	9.8 d	3952 cd	5210 de	0.0a	5.7ab	2.7a	3.0ab
R.S.+F.S twice	0.2%+0.1%	5.0	23.5 de	43.8 e	2830 e	3128 f	0.0a	0.7a	10.7a	1.3a
R.C	2.0%	9.5	29.9 e	72.9 f		2666 f	8.7 b	8.0b	76.0b	107.3c
Control	-	-	88.7 f	97.1 g		1397 g	18.0c	9.3b	106.0b	108.7c

In a column the values followed by the same letter do not differ significantly according to Duncan's multiple range test (P+0.05).

a = average values of three replications.
R.Z = Root zone,
B.C. = Broadast,
R.C = Root Coat.

S.I = Soil incorportion,
F.S. = Foliar spray,
R.S. = Root Soak,

Table 9 : Disease incidence, grain yield, and leafhopper population of Taichung Native 1 (T) and Ratna (R)[a] rice treated with insecticides (Satapathy and Anjaneyulu, 1982)

Treatment	*Disease incidence (%)*		*Grain yield (t/ha)*		*Leafhoppers (No,/20 hills)a*			
					Adults		*Nymphs*	
	T	*R*	*T*	*R*	*T*	*R*	*T*	*R*
Cypermethrin	3.0a	0.3a	5.3a	6.6a	0.0a	0.0a	0.0a	0.0a
FMC 35001	50.8c	3.4b	1.4c	5.0b	13.7c	13.7b	0.7a	1.3ab
Phosphomidon	56.7d	6.1c	1.1de	4.4bcd	38.7cd	14.7b	1.3ab	1.0ab
Demeton.0. methyl sulphoxide	58.1d	4.5bc	1.2cd	4.5bc	40.0cd	16.0bc	6.0bc	0.0a
Ofunack	71.5e	12.5d	0.9ef	4.0cd	42.3d	18.7c	8.7c	0.7ab
Dichlorvos	73.1e	11.2d	0.9f	3.6d	46.0d	23.0d	33.3d	2.7ab
Acephate	31.2b	2.8b	1.8b	5.0b	22.7b	15.3bc	0.7a	0.7ab
Control	100.0f	51.5e	0.3g	2.7e	68.7e	41.3e	48.7e	12.0c

a - Values followed by a common letter do not differe significantly by Duncan's Multiple range test (P= 0.05).

b - Av values of 3 replications.

References

Abenes, M.L.P. and Z.R. Khan 1990. Attration of rice leafhoppers and planthoppers to different light colours. *Int. Rice Res. Newsl.* 15(2): 32-33.

Ali. Md. A and S.A. Miah. 1990. Presence of rice tungro viruses in the first ratoon crop of tungro infected main crop of rice. *Bangladesh J. Bot.* 19: 155-157.

Anjaneyulu, A. 1974. Epidemiological studies on rice tungro virus in India. International Rice Research Conference. April 22-26, 1974. IRRI, Philippines, p. 12.

— A. 1975a. *Nephotettix virescens* (Distant) nymphs and their role in the spread of rice tungro virus. *Curr. Sci.* 44: 357-358.

— A. 1975b. A method of inducing rice tungro virus epiphytotic. *Int. Rice Res. Newsl.* 24: 107-108.

Anjaneyulu, A., V.D. Shukla, G.M. Rao and S.K. Singh. 1981. Perpetuation of rice tungro virus and its vectors. *Int. Rice Res. Newsl.* 6: 12-13.

Anjaneyulu, A., S.K. Singh and M.M. Shenoi, 1982a. Evaluation of rice varieties for tungro resistance by field screening technique. *Trop. Pest Managm.* 28: 147-156.

Anjaneyulu, A., V.D. Shukla, G.M. Rao and S.K. Singh, 1982b. Experimental host range of rice tungro virus and its vectors. *Plant Dis.* 66: 54-56.

Anjaneyulu, A., M.K. Satapathy, and V.D. Shukla (1999) *Rice Tungro. Science* Publishers. Inc. Lebanon, NH 03766, USA. p. 228.

Anjaneyulu, A., R.D. Daquioag, M.E. Mesina, H. Bibino, R.T. Lubigan and K. Moody. 1988. Host plants of rice tungro (RTV) - associated viruses. Int. Rice Res. Newsl. 13(4): 30-31.

Anjaneyulu, A., V.D. Shukla, G.M. Rao and S.K. Singh. 1981. Perpetuation of rice tungro virus and its vectors. *Int. Rie Res. Newsl.* 6: 12-13.

Anjaneyulu, A. and N.K. Chakrabarti, 1977. Geographical distribution of rice tungro virus disease and its vectors in India. *Int. Rice Res. Newsl.* 2(5): 15-16.

Astika, N.S., N. Saweta, G.N. Aryawan and Y. Suzuki (1992) Dependence of incubation period and symptoms of rice tungro disease (RTD) in infection stage in rice fields. *Int. Rice Res. Newsl.* 17 (13): 19-20.

Azzam, O, S.W. Ahn, G. Khush, G.L. Raina, S.S. Cheema and G.S. Sidhu 1999. Outbreak of yellow stunt syndrome on rice in Punjab, India. *IRRI Res Notes* 24(1): 43-44.

Bajet, N.B., V.M. Aguiero, R.D. Daquioag, G.B. Jonson, R.C. Cabunagan, E.M. Mesina and H. Hibino. 1986. Occurrence and spread of rice tungro spherical virus in the Philippines. *Plant Dis.* 70: 971-973.

Bejet, N.B., R.D. Daquioag and H. Hibino. 1985. Enzyme linked immunosorbent assay to diagnose rice tungro. *J. Pl. Prot. Tropics.* 2: 125-129.

Baria, A.R. (1997) Status of rice tungro disease in the Philippines: a guide to current and future research. In *Epidemiology and Management of rice tungro disease* (Ed. T.C.B. Chancellor and J.M. Thresh). Natural Resources Institute, UK, pp. 76-83.

Berger, R.D. 1988. The analysis of control measures on the development of epidemics. In: *Experimental Techniques in Plant Disease Epidemiology* (Ed. by J. Kranz and J. Rotem) pp. 137-151. Springerverlag, Berlin.

Bottenberg, H. and J.A. Litsinger. 1989. Using fluorescent dye to map dispersal pattern of rice green leafhopper (GLH) *Int. Rice Res. Newsl.* 14(6): 25.

Bottenberg, H., J.A. Litsinger, A.T. Barrion and P.E. Kenmore. 1990. Presence of tungro vectors and their natural enemies in different rice habitats in Malaysia. *Agric. Ecosystems Environ.* 31: 1-15.

Cabauatan, P.Q., N. Kobayashi, R. Ikeda and H. koganezawa (1993). *Oryza glaberrima*: an indicator plant for rice tungro spherical virus. *Intl. J. Pest. Mgmt.* 39: 273-276.

Cabauatan, P.Q., C.R. Cabunagan and H. Koganezawa (1995) Biological variants of rice tungro viruses in the Philippines. *Phytopathology* 85: 77-81.

Cabauatan, P.Q. and H. Hibino. 1985. Transmission of rice tungro bacilliform and spherical viruses by *Nehpotettix virescens* Distant. *Philippine Phytopath* 21: 103-109.

Cabautan, P.Q. and H. Hibino. 1988. Isolation, purification and serology of rice tungro bacilliform and rice tungro spherical viruses. *Plant Dis.* 72: 526-528.

Cabunagan, R. C., Z.M. Flores, E.R. Tiongco and H. Hibinio, 1987. Diagnostic techniques for rice tungro disease pp. 13-17. In *Proceedings of the workshop on Rice Tungro virus.* Ministry of Agriculture. AARD-Maros Research Institute for food crops. Maros, Indonesia.

Carbonell, M.P. and K.C. Ling. 1971. Transmission of tungro by nymphs of Nephotettix impicticeps. Philippine Phytopath. 7: 1-2 (abstract).

Carino, F.O. 1981. Role of natural enemies in population suppression and pest management of green rice leafhoppers. Ph. D. thesis. University of the Philippines at Los Banos Laguna, p. 136.

Chancellor, T.C.B. and A.G. Cook (1995). The bionomics and population dynamics of *Nephotettix* spp. (Hemipteran: (Cicadellidae) in South and Southeast Asia with particular reference to the incidence of rice tungro virus disease. *Trop. Sci.* 35: 200-216.

Chancellor, T.C.B. K.L. Heong, A.G. Cook and S. Villareal (1996a) Leafhopper ecology and rice tungro disease dynamics. In rice tungro disease epidemiology and vector ecology: the development of sustainable and cost effective pest management practices to reduce yield losses in intensive rice cropping systems (Ed. by T.C.B. Chancellor, P.S. Teng and K.L. Heong) IRRI Discussion paper series no. 19, IRRI, 933 Mamila, Philippines. pp. 10-35.

Chancellor, T.C.B., A.G. Cook and K.L. Heong (1996a). The within-field dynamics of rice tungro disease in relation to the abundance of its major leafhopper vectors. Crop protection 15(5): 439-449.

Chancellor, T.C.B., E.R. Tiongco, J. Holt, S. Villareal P.S. Teng, N. Fabellor, and M.G. Magbama (1996b). Risk. Factors of rice tungro disease in endemic areas. In rice tungro disease epidemiology and vector ecology: the development of sustainable and cost effective pest management practices to reduce yield losses in intensive rice cropping systems (Ed. by T.C.B. Chancellor, P.S. Teng and K.L. Heong) IRRI discussion paper series no. 19, IRRI, 933 Mamila, Philippines. pp. 58-73.

Chancellor, T.C.B. and J.M. Thresh (eds) (1997) *Epidemiology and Management of Rice Tungro Disease.* Chatham, UK, Natural Resources Institute, p. 108.

Chen, Y.M. and A.T. Jatil (1997). Incidence and control of rice tungro disease in Malaysia pp. 103-108. In *Epidemiology and management of rice tungro disease* (Ed. by Chancellor, T.C.B. and J.M. Thresh) Natural Resources Institute, Chattam, UK.

Chen, Y.M. and A.B. Othman. 1991. Tungro in Malaysia. *MAPPS Newsal.* 15(1): 5-6.

Cheng. C.H. and M.D. Pathak. 1972. Resistance to *Nephotettix virescens* in rice varieties. *J. Econ. Entomol.* 15: 1-16.

Chowdhary, A.K. (1997) Present status of rice tungro disease in India. In *Epidemiology and Management of rice tungro disease* (Ed: T.C.B. Chancellor and J.M. Thresh), Natural Research Institute, UK pp. 69-75.

Cook, A.G. and T.J. Perfect. 1989. Population dynamics of three leafhopper vectors of rice tungro viruses *Nephotettix virescens* (Distant), *N. nigropictus* (Stal) and *Recilia doraslis* (Motschulsky) (Hemiptera: Cicadellidae) in farmers fields in the philippines. *Bull. Ent. Res.* 79:437-451.

Cooter, R.J. and D. Winder (1996) The flight behaviour and dispersal range of leafhopper vectors of rice tungro disease. *In Rice tungro disease epidemiology and vector ecology*: the development of sustainable and cost effective pest management practices to reduce yield losses in intensive rice cropping systems (Ed. by T.C.B. Chancellor, P.S. Teng and K.C. Heeng) IRRI Discussion paper series no. 19, IRRI, Post Box 933 Manila, Philippines.

Dahal, G., H. Hibino, R.C. Cabunagan, E.R. Tiongco, Z.M. Flores and V.M. Aquiero. 1990a. Changes in cultivar reactions to tungro due to changes in virulence of the leafhopper vector. *Phytopathology* 80: 659-665.

Dahal, G.,R.B. Shrestha, N.K. Khatri, Z. Fan and R. Hull (1993) Incidence of virus disease of rice in Nepal. *Journal of the Institute of Agriculture and Animal Sciences* 14: 115-116.

Dahal, G. (1997) Rice Tungo Disease and green leafhopper vector in Nepal: Current Status and future research strategies. In *Epidemiology and Management of rice tungro disease (Ed:* I.C.B. Chancellor and J.M. Threst) Natural Resources Institute, UK, pp. 84-93.

Dahal, G. H. Hibino and R.C. Sexena. 1990b. Association of leafhopper feeding behaviour with transmission of rice tungro to susceptible and resistant rice cultivars *Phytopathology* 80:371-377.

Dahal, G., V.M. Aquiero, R.C. Cabunagan and H. Hibino. 1988. Varietal reaction to tungro (RTV) with changes in leafhopper virulence. *Int. Rice Res. Newsl.* 13(5): 12-13.

Dahal, G. and H. Hibino. 1985. Varieties with different resistance to tungro (RTV) and green leafhopper (GLH). *Int. rice Res. Newsl.* 10(1): 5-6.

Daquioag, R.D., E.R. Tiongco and H. Hibnino. 1984. Reaction of several rice varieties to rice tungro virus (RTV) complex *int. Rice Res. Newsl.* 9(2): 5-6.

Furuta, T. 1977. Rice Waika, a new virus disease, found in Kyushu, Japan, *Rev. Plant Protect. Res.* 10:70-82.

Gupta, M.G., A.K. Singh, S.F. Hameed and D.N. Mehto. 199. Seasonal prevalence of green leafhopper on rice in North Bihar. *Oryza* 26:411-413.

Habibuddin, H., T. Takita and N.K. Ho. 1987. Research and management of tungro disease in peninsular Malaysia. pp. 86-91. In: *Proceedings of the workshop on Rice Tungro virus.* Ministry of Agriculture, AARD Maros Research Institute for food crops. Maros, Indonesia.

Hasanuddin, A. and K.C. Ling. 1980. Effect of different tungro infected varieties as virus sources on the infectivity of *Nephotettix virescens. Int. Rice. Res Newsl.* 5(1): 5.

Hasanuddin, A. and H. Hibino. 1989. Grain yield reduction growth retardation, and virus concentration in rice plants infected with tungro-associated viruses. pp 56-73. *Crop losses due to disease outbreaks in the tropics and counter measures. Trop. Agric. Res. Ser.* No. 22. Tropical Agricultural Research Center, Tsukuba; Japan.

Hasanuddin, A., Kosesnang and D. Baco (1997) Rice Tungo virus disease in Indonesia: Present status and current management strategy pp. 84-93. In *Epidemiology and management of rice tungro disease* (Ed. by Chancellor, T.C.B and J.M. Thresh) Natural Resources Institute, Chatham, UK.

Hay, J.M., M.C. Jones, M.L. Blankebrough, I. Dasgupta, J.W. Davies and R. Hull. 1991. An analysis of the sequence of an infectious clone of rice tungro bacilliform virus a pararetro virus *Nucleic Acids Res.* 19: 2615-2621.

Heinrichs, E.A. and H. Rapusas. 1984. Feeding, development and tungro virus transmission by the green leafhopper, *Nephotettix virescence* (Distant) Homoptera: Cicadellidae) after selection on resistant rice cultivars. *Environ. Entomol* 13: 1074-1078.

Heinrichs. E.A. 1979. Control of leafhopper and planthopper vectors of rice viruses. pp. 529-558. *In: Leafhopper Vectors and Plant Disease Agents.* (Ed. by: K. Maramorosch and K.F. Harris). Academic Press Inc., New York, p. 654.

Herdt, R.W. 1988. Equity considerations in setting priorities for third world rice biotechnology research. *Development: Seeds of change* 4: 19-24.

Hibino, H. and A. Anjaneyulu. 1991. Towards stable resistance to rice virus diseases. *In International Plant Protection: Focus on the developing World. Proc. 11th Int. Cong. Plant Protection*, Vol.2, (Ed. E.D. Magallona) 5-9 Oct.1987, Manali, Philippines. pp. 27-31.

Hibino, H. N.,Saleh and M. Moechan 1979. Transmission of two kinds of rice tungro-associated viruses by insect vectors. *Phytopathology* 69: 1266-1268.

Hibino, H., M Roechan and S. Sudarisman. 1978a. Association of two types of viruses particles with penyakit habang (tungro disease) of rice in Indonesia. *Phytopathology* 68: 1412-1416.

Hibino. H. 1987. Rice Tungro virus disease: current research and future prospects. pp. 2-6. In *Proceeding of the workshop on rice tungro virus.* Ministry of Agriculture. AARD-Maros Research Institute for food crops. Maros Indonesia.

Hibino, H. 1989. Insect-borne viruses of rice. pp. 209-241 In *Advances in Disease vector Research.* vol.6 (Ed. by: Harris, K.F.), springler-verlag, New York.

Hibino, H., R.D. Daquioag, P.Q. Cabauatan and G. Dahal. 1988. Resistance to rice tungro spherical virus in rice. *Plant Dis.* 72: 843-847.

Hibino. H., K. Ishikawa, T. Omura, P.Q. Cabauatan and H. Koganezawa 1991. Characterization of rice tungro bacilliform and rice tungro spherical viruses. *Phytopathology* 81: 1130-1132.

Hibino. H., R.D. Daquioag, E.M. Mesina and V.M. Aquiero. 1990. Resistance in rice tot tungro-associated viruses *Plant Dis.* 74: 923-926.

Hibino. H., N. Saleh, H. Jumanto, S. Sudarisman and M. Roechan. 1978b. Two kinds of virus particles associated with tungro disease of rice. *Ann. Phytopath. Soc. Japan* 44:- 394 (In Japanese).

Hino. T., L Watanakul, M. Nabheerong, P. Surin, W. Chaimongkal, S. Disthaporn, M. Putta, D. Kerdchokchai and A. Surin. 1974 studies on rice yellow-orange leaf virus disease in Thailand. Tech. Bull. No.7. Tropical Agriculture Research Centre. p. 67

Holt, J. and T.C.B. Chancellor (1997). A model of plant virus disease epidemics in a synchronously planted cropping systems. *Plant Pathology* 46: 490-501.

Holt, J. and T.C.B. Chancellor (1996b) Simulation modelling of the spread of rice tungro virus disease; the potential for management by roguing. *J. Appl.* Ecol. 33: 927-936.

Holt, J., T.C.B. Chancellor and M.K. Satapathy, 1992. A prototype simulation model to explore options for the management of rice tungro virus disease. pp. 973-980. In *Proceedings of Brighton crop protection conference-Pests and diseases.* 23-26 Nov. 1992. Brighton, U.K.

Holt. J. (1996). Spatial modelling of rice tungro disease epidemics. In rice tungro disease epidemiology and vector ecology: the development of sustainable and cost effective pest management practices to reduce yield losses in intensive rice cropping systems (Ed. by T.C.B. Chancellor, P.S. Teng and K.L. Heong) IRRI Discussion paper series no. 19, IRRI, 933 Mamila, Philippines. pp. 74-85.

Hull, R., M.C. Jones, I. Dasgupta, J.M. Cliffe, C. Mingins, G. Lee and J.W. Davies. 1991. Molecular biology of rice tungro viruses: evidence for a new retroid virus. pp. 539-547. In *Rice Genetics II, International* Rice Research Institute, Los Baños, Philippines. p. 844.

Hull, R., J.M. Hay, A. Druka Z. Fan, V. Thole, and S. Zhang 1997) Molecular biology of variation in rice tungro viruses. In rice tungro disease epidemiology and vector ecology: the development of sustainable and cost effective pest management practices to reduce yield losses in intensive rice cropping systems (Ed. by T.C.B. Chancellor, P.S. Teng and K.L. Heong) IRRI Discussion paper series no. 19, IRRI, 933 Mamila, Philippines. pp. 5-10.

Inoue, H.S. Ruay-aree, C. Hengsawad, V. Hengsaward and P. Patirupanusara. 1975. Studies on rice green leafhopper in Thailand in relation to yellow orange leaf virus disease. Final report of the joint research work between Thailand and Japan. p. 140.

IRRI 1973, International Rice Research Institute. Annual Report for 1972. Los Banos, Philippines, p. 246.

IRRI 1983, International Rice Research Institute. Annual Report for 1982. Los Banos, Philippines, p. 532.

IRRI 1984. International Rice Research Institute, Annual Report for 1983. Los Banos, Philippines, p. 494.

IRRI 1985. International Rice Research: 25 years of Partnership. International Rice Research Institute. Los Banos, Philippines, p. 188.

IRRI 1992. International Rice Research Institute, Program Report for 1991, p. 322.

IRRI 1987. International Rice Research Institute, Annual Report for 1986 Los Banos, Philippines, p. 639.

John, V.T. 1968. Identification and charaterization of tungro, a virus disease of rice in India. *Plant Dis. Reptr.* 52: 871-875.

John, V.T. and A Ghosh, 1981. Estimation of yield losses due to rice tungro virus. *Indian J. Agric. Sci.* 51: 48-50.

Jhon, V.T., W.H. Freeman and B.B. Shahi. 1979. Occurrence of tungro disease in Nepal. *Int. Rice Res. Newsl.* 4(5): 76.

Jones, M.C., K. Gough, I. Dasgupta, B.L. Subba Rao, J. Cliffe, R. Qu, P. Shen; M. Kaniewska, M. Blakebrough, J.W. Davies, R.N. Beachy and R. Hull. 1991. Rice tungro disease is caused by an RNA and a DNA virus. *J. Gen. virol.* 72: 757-761.

Kano, H., M. Koizumi, H. Noda, H. Hibino, K. Ishikawa, T. Omura, P.Q. Cabauatan and H. Koganezawa. 1992. Nucleotide sequence of capsid protein gene of rice tungro baciliform virus. *Arch. Virol.* 124: 157-163.

Khan, M.A., H. Hibino, V.M. Aquiero and R.D. Daquioag. 1991. Rice and weed hosts of rice tungro-associated viruses and leafhopper vectors. *Plant Dis.* 75: 926-930.

Khan, Z.R. and R.C. Saxena. 1985. Behaviour and biology of *Nephotettix virescens* (Homoptera: Cicadellidae) on tungro virus infected rice plants: Epidemiology implications. *Environ. Entomol* 14: 297-304.

Kiritani, K., N. Hokyo, T. Sasaba and F. Nakasuji. 1970. Studies on population dynamics of the rice green leafhopper. *Nephotettix cincticeps Uhler*: regulatory mechanism of the population density. *Res popul. Ecol.* 12: 137-153.

Kiritani, K. (1981) Spacio-temporal aspects of Epidemiology in Insect borne rice virus diseases *JAR* 15(2): 92-99.

Koganezawa, H. and R.C. Cabunagan (1997). Resistance to rice tungro virus disease pp. 54-59. In Epidemiology and Management of rice tungro disease (Ed. by Chancellor, T.C.B. and J.M. Thresh) Natural Resources Institute, Chatham, UK.

Kondaiah, A., A.V. Rao and T.E. Srinivasan. 1976a. Factors favouring spread of rice 'tungro, disease under field conditions. *Plant dis. Reptr.* 60: 803-806.

Kondaiah, A., V.T. John and A.P.K. Reddy. 1976b. *In vitro* isolation of rice tungro virus from cut leaves of rice. *Int. Rice Res. newl.* 1(1): 18.

Lamey, H.A., P. Surin and J. Leeuwangh. 1967. Transmission experiments on the tungro virus in Thailand. *Int. Rice Comm. Newsl.* 16: 15-19.

Lim, G.S. 1972. Studies on penyakit merah disease of rice. III. Factors contributing to an epidemic in North Krian, Malaysia. *Malaysia Agric. J.* 48: 278-294.

Lim, G.S., W.P. Ting and K.L. Heong. 1974. Epidemiological studies on tungro virus in Malaysia. Malaysian Agricultural Research and Development Institute (MARDI) Report p. 21,12.

Ling, K.C. 1976. Recent studies on rice tungro disease at IRRI. *IRRI Res. Pap. Ser. no* 1, p. 11.

Ling, K.C. 1975a. Experimental epidemiology of rice tungro disease. 1. Effect of some factors of vector (*Nephotettix virescens*) on disease incidence. *Philippine Phyopath.* 11: 11-20.

—1975b. Experimental epidemiology of rice tungro disease. II. Effect of virus source on disease incidence. *Philippine Phytopath.* 11: 21-31.

Ling, K.C. and M.K. Palomar. 1966. Studies on rice plants infected with the tungro virus at different ages. *Philippine Agric.* 50: 165-177.

Ling, K.C. and E.R. Tiongco. 1977. Transmission of rice tungro virus at various temperatures : A transitory virus-vector interaction. *IRRI Res. Pap Ser.* 4, p. 26.

Ling, K.C. 1968. Hybrids of *Nephotettix impicticeps* Ishihara and *N. apicalis* (Motsch). and their ability to transmit the tungro virus of rice. *Bull. Entomol. Res.* 58: 393-398.

— 1969. Non-propagative leafhopper-borne viruses. pp. 255-277. In *Viruses, Vector and Vegetation.* (Ed. by K. Maramorosch). Interscience Publishers, New York, p. 666.

— 1972. Rice virus disease. *Int. Rice Res. Inst, Los* Banon, Philippines. p. 142.

— 1970. Ability of *Nephotettix apicalis* to transmit the rice tungro virus. *J. Econ. Entomol.* 63: 583-586.

— 1966. Non-persistence of the tungro virus of rice in its leafhopper vector, *Nephotettix impicticeps. Phytopathology* 56: 1252-1256.

Litsinger, J.A. (1989) Second generation insect pest problem on high yielding rices *Trop. Pest Mgmt.* 35: 235-242.

Lowe, J.A. and P. Nandi. 1972. Surveillance of pests and diseases of rice in India with special reference to the occurrence of tungro virus (mimeo) IRRI seminar, p. 4.

Mallick, S.C., A.K. Chowdhary and D. Pal (1999). *Ischaemum rugosum*- a potential alternate host of rice tungro viruses in West Bengal. IRRI Res. Notes. 24(1): 28-29.

Manwan, I., S. Sama and S.A. Rizvi. 1985. Use of varietal rotation in the management of tungro disease in Indonesia. *Indonesian Agric. Res. Dev. J.* 7: 43-48.

Mariappan. V. and R.C. Saxena. 1983a. Effect of blends of custard-apple oil and neem oil on survival of *Nephotettix virescens* and tungro virus transmission. *Int. Rice Res. Newsl.* 8(4): 15-16.

Miah, S.A. and K.C. Ling. 1978. Effect of temperature on the movement of *Nephotettix virescens. Int. Rice Res. Newsl.* 3(5): 13.

Mishra, M.D., A. Ghosh, F.R. Niazi, A.N. Basu and S.P. Rychaudhuri. 1973. The role of graminaceous weeds in the perpetuation of rice tungro virus. *J. Indian Bot. Soc.* 52: 176-183.

Mishra, MD., A. Ghosh, F.R. Niazi, A.N. Basu and S.P. Raychaudhuri. 1973. The role of graminaceous weeds in the perpetuation of rice tungro virus. *J. Indian Bot. Soc.* 56: 176-183.

Mochida, O., S.L. Valencia and R.P. Basilio. 1986. Chemical control of green leafhoppers to prevent virus diseases, especially tungro disease on susceptible/intermediate rice cultivars in the tropics. *Trop. Agric. Res. Ser.* No. 19. Tropical Agriculture Research Centre, Tsukuba, Japan, pp. 195-208.

Mohanty S.K., G. Bhaktavatsalam and S.K. Singh. 1987. A new weed host of rice tungro complex. *Curr. Sci.* 56: 1185-1186.

Mohanty, S.K., M.K. Satapathy and R. Sridhar. 1983. Significance of virus induced proline accumulation in rice tungro epidemics. *Curr. Sci.* 52: 311-314.

Mukhopadhyay, S. 1986. Ecosystem analysis of rice green leafhoppers and rice tungro virus epidemiology in West Bengal, India. Paper presented at Rice Tungro Virus Technical Meeting. 24-27 Sept., 1986. Maros, Indonesia. p. 33.

Mukhopadhyay, S., A.B. Ghosh, P. Tarafder and S. Chakravarti. 1978. Studies on rice tungro virus diseases and its vector, *nephotettix* spp. in West Bengal, India. *Int. Rice Res. Newsl.* 3(4): 14.

Mukhopadhyay, S. and A.K. Chowdhury, 1973. Some epidemiological aspects of tungro virus disease of rice in West Bengal. *Int. Rice Comm. Newsl.* 22: 44-57.

Mukhopadhyay, S. and A.K. Chowdhury, 1970. Incidence of tungro virus of rice in West Bengal. *Int. Rice Comm. Newsl.* 19(2): 9-12.

Mukhopadhyay, S. 1980. Ecology of *Nephotettix* spp. and its reaction with rice tungro virus. Final Report, Indian Council of Agricultural Research, New Delhi.

Mukhopadhyay, S. and S. Mukhopadhyay. 1986. Prediction of the peak appearance of rice green leafhoppers in West Bengal. pp. 22-25. In *proceeding of the Workshop on Epidemiology of plant virus diseases.* 6-8 Aug., 1986, Corlardo, Florida, U.S.A.

Narayanasamy, P. 1972. Influence of age of the rice plant at the time of inoculation on the recovery of tungro virus by *Nephotettix impicticeps* (Ishihara). *Phytopathol.* Z. 74: 109-144.

Oka, I.N. 1971. On an outbreak of a rice disease showing tungro symptoms in South Kalimantan, South Sumatra and Lampung Provinces. Central Research Institute for Agriculture, Bogor, Indonesia, 4p. (Mimeo).

Omura, T., H. Inoue, U.B. Thapa and Y. Saito. 1981. Association of rice tungro spherical and rice tungro bacilliform virus with the disease in Jankpur, Nepal. *Int. Rice res. Newsl.* 6(6): 14.

Ou, S.H. 1965. Rice diseases of obscure nature in tropical Asia with special reference to 'mentek' disease in Indonesia. *Int. Rice Comm. Newsl.* 14(2): 4-10.

Ou, S.H. and K.G. Goh. 1966. Further experiment of "Penyakit merah" disease of rice in Malaysia. *Int. Rice Comm. Newsl.* 15: 31-33.

Ou, S.H. K.C. Ling and H.E. Kauffman. 1974. Current status of disease control. International Rice Research Conference, April 22-26, 1974, IRRI, Philippines. p. 15.

Palmer, L.T. and P.S Rao, 1981. Grassy stunt, ragged stunt and tungro diseases of rice in Indonesia. *Trop Pest Managm.* 27: 212-217.

Parejarearn, A.,D. Chettanchit, M. Putta, W. Rattanakarn, J. Arayapan and S. Disthaporn. 1990. Hosts of rice tungro-associated viruses (RTVs) in Thailand. *Int. Rice Res. Newsl.* 15(6): 21-22.

Perfect, T.J. and A.G. Cook. 1982. Diurnal periodicity of flight in some Delphacidae and Cicadellidae associated with rice. *J. Eco. Entomol.* 7: 317-326.

Peries, I.D.R., G.A.W. Wijesekera and C.M. Bandaranayaka. 1985. Survey of tungro and other virus disease in the central region of Sri Lanka: *Trop. Agriculturist* 141: 121-122.

QU.R, M. Bhattacharyya, G.S. Laco, A. Kochko, De, B.L. Subba Rao, M.B. Kaniewska, J.S. Elmer, D.E. Rochester, C.E. Smith and R.N. Beachy (1991). Characterization of the genome of rice tungro bacilliform virus: Comparison with commelina yellow mottle virus and caulimoviruses. Virology 185: 354-364.

Rahman. M.M., M.A. Nahar and S.A. Miah. 1985. Tungro spread in rice fields. *Int. Rice Res. Newsl.* 10(6): 18.

Rao, G.M. and A. Anjaneyulu. 1980. Estimation of yield losses due to tungro virus infection in rice cultivars. *Oryza* 17: 210-214.

Rao, G.M. and A. Anjaneyulu. 1982. Effect of meteorological factors on symptomatology and acquisition of rice tungro virus by *Nephotettix virescens. Int. Rice Res. Newsl.* 7(6): 9.

Rao, P.S. and A. Hasanuddin. 1991. Incidence of rice tungro virus disease and its vector in South Sulawesi, Indonesia. *Trop. Pest. Managm.* 37: 256-258.

Rao, G.M. 1977. Virus-vector-host relationships of rice tungro virus. Ph.D. thesis, Utkal University, Bhubaneswar. p. 170.

Rao, G.M. and A. Anjaneyulu. 1979. Rice tungro virus acquisition by the vector, *Nephotettix* virescens from rice cultivars. *Plant Dis. Reptr.* 63: 855-858.

Rapusas, H.R. and E.A. Heinrichs. 1987. Plant age effect on the level of resistance of rice "IR36" to the green leafhopper, *Nephotettix virescens* (Distant) and rice tungro virus. *Environ. Entomol.* 16: 106-110.

Raychaudhuri, S.P., M.D. Mishra and A. Ghosh. 1967. Preliminary rote on transmission of a virus disease resembling tungro of rice in India and other virus like symptoms. *Plant Dis. Reptr.* 51: 300-301.

Raychaudhuri, S.P. M.D. Mishra and A.N. Basu. 1971. Investigations on tungro and yellow dwarf diseases of rice. *Oryza* 8(2): 316-363.

Reddy, D.B. 1973. High yielding varieties and special plant protection problems with particular reference to tungro virus of rice *Int. Rice Comm. Newsl.* 22: 34-42.

Reyes, T.T. 1989. Selected economically important diseases of some major crop in the Philippines. pp. 11-20 *Tropical Agriculture Research Series* No. 22. Tropical Agricultural Research Centre. Tsukuba,Japan.

Riley, J.R., D.R. Reynolds and R.A. Farrow. 1987. The migration of *Nilaparvata lugens* (Stal) (Delphacidae) and other Hemiptera associated with rice during the dry season in the Philippines: a study using radar, visual observations, aerial nettings and ground trapping. *Bull. Ent. Res.* 77: 145-169.

Rivera, C.T. and S.H. Ou. 1965. Leafhopper transmission of tungro disease of rice. *Plant Dis. Reptr.* 49: 127-131.

Rivera, C.T., K.C. Long and S.H. Ou and V.M. Aquiero. 1969. Transmission of two strains of rice tungro virus by *Recillia dorsalis. Philippine Phytopathol.* 5: 17 (Abstr.)

Saikia, A.K., K.N. Bhagabati, Y. Rathaiah and H.H. Choudhury. 1992. Rice tungro disease (RTD) incidence in Assam, India. *Int. Rice Res. Newsl.* 17(6): 25.

Saito, Y. 1977. Inter-relationship among waika disease, tungro and other similar diseases of rice in Asia. pp 129-135. *Trop. Agric. Res. Ser. No.* 10. Tropical Agriculture Research Center, Tsukuba, Ibaraki, Japan.

Satapathy, M.K. and A. Anjaneyulu. 1982. Insecticidal control of rice tungro virus disease. *Int. Rice. Res. Newsl.* 7(6): 9-10.

Satapathy, M.K. and A. Anjaneyulu. 1984. Cypermethrin, a synthetic pyrethroid in the control of rice tungro virus disease and its vector. *Trop. Pest Managm.* 30: 170-178.

Satapathy, M.K. and A. Anjaneyulu. 1989a. Effect of root zone placement of granular insecticides for tungro prevention and its control. *Trop. Pest Managm.* 35: 51-56.

Satapathy, M.K. and A. Anjaneyulu, 1989b. Evaluation of wettable powder insecticides for tungro prevention and its management. *Trop. Pest Managm.* 35: 41-47.

Satapathy, M.K. and A. Anjaneyulu, 1991. Rice tungro management by different methods of carbofuran application. *Oryza* 28: 377-387.

Satapathy, M.K. and A. Anjaneyulu. 1983. Evaluation of emulsifiable concentrate and wettable powder insecticides for control of tungro virus disease and its vector, *J. Pl. Dis. Protect.* 90: 269-277.

Satapathy, M.K. and A. Anjaneyulu. 1986. Prevention of rice tungro virus disease and control of the vector with granular insecticides. *Ann. Appl. Biol.* 108: 503-510.

Satapathy, M.K. and A. Anjaneyulu. 1990. Carbofuran as root zone and broadcast application for tungro control- a comparative study. *Int. J. Trop. Pl. Dis.* 8: 65-67.

Satapathy, M.K. and A. Anjaneyulu. 1989c. Experimental epidemics of tungro and its vectors in nursery beds under different pesticide treatments. *Int. J. Trop. Pl. Dis.* 7: 137-150.

Satapathy, M.K., T.C.B. Chancellor, E.R. Tiongco, P.S. Teng and J.M. Thresh (1996). The effect of internal and external inoculum sources on tungro disease spread. In rice tungro disease epidemiology and vector ecology: the development of sustainable and cost effective pest management practices to reduce yield losses in intensive rice cropping systems (Ed. by T.C.B. Chancellor, P.S. Teng and K.L. Heong) IRRI Discussion paper series no. 19, IRRI, 933 Mamila, Philippines. pp. 36-57.

Satapathy, M.K. (1998) Chemical control of insect and nematode vectors of plant viruses. pp. 188-195. In *Plant virus disease control* (Ed. by A. Hadidi; R.K. Khetarpal and H. Koganezawa). APS press, USA.

Satapathy, M.K., P.S. Teng and A. Anjaneyulu (1998) Rice tungro disease and its management. pp. 105-130. In *IPM system in Agriculture* vol. 3. (Ed. by R.K. Upadhyay, K.G. Mukharjee and R.J. Rajak) Aditya Books Pvt.Ltd., New Delhi, India.

Savary, S., N. Fabellar, E.R. longco and P.S. Teng (1993). A characterization of rice tungro epidemics in the Philippines from historical survey data. *Plant Dis.* 77: 376-382.

Sexena, R.C. and H.D. Justo, 1986. Effect of custard apple oil, neem oil and neem cake on green leafhopper (GLH) population and on tungro (RTV) infection. *Int. Rice Res. Newsl.* 11(2): 25.

Serrano, F.B. 1957. Rice 'accepna pula' or stunt disease- a serious meanace to the Philippine rice industry. *Philippine J. Sci.* 86: 203-230.

Shen, P., M. Kaniewska, C. Smith and R.N. Beady. 1993. Nucleotide sequence and genomic organisation of rice tungro spherical virus. Virology 193: 621-630.

Shinkai, A. 1971. Rice waika, a new virus disease found in Kyushu, Japan. pp. 123-127. *Trop. Agric. Res. Ser.* 10 Tropical Agriculture Research Center, Tsukuba, Japan.

Shukla, V.D. 1979. Studies on the epidemiology of rice tungro virus disease. Ph.D. thesis, Utkal Univ., Bhubaneswar. p. 249.

Shukla, V.D. and A. Anjaneyulu. 1980. Evaluation of systematic insecticides for control of rice tungro. *Pl. dis.* 64: 790-792.

Shukla, V.D. and A. Anjaneyulu. 1981. Adjustment of planting date to reduce rice tungro disease. *Plant Dis.* 65: 409-411.

Shukla, V.D. and A. Anjaneyulu. 1982a. Effect of number of leafhoppers and amount and source of virus inoculum on the spread of rice tungro. *J. Pl. Dis. Protect.* 89: 325-331.

Shukla, V.D. and A. Anjaneyulu. 1982b. Spread of tungro virus disease in irrigated and rainfed rice. *Indian Phytopath.* 35: 47-51.

Shukla, V.D. and A. Anjaneyulu. 1982c. Population and sex ratios of rice green leafhoppers in the paddy field. *Sci. and Cult.* 48: 65-66.

Singh, K.G. 1967. Penyakit merah disease, a virus infection of rice in Malaysia. pp. 75-80. In Proceedings of symposium on the virus diseases of rice plants. 35-28 April, 1967, Los Banos, Philippines.

Singh, K.G. 1971. Studies on Transmission of penyakit merah virus disease of rice. *Malaysian Agric. J.* 48: 93-103.

Singh, S.K., M.M. Shenoi and A. Anjaneyulu. 1982. Tungro incidence in Bihar and West Bengal of India during 1981. *Int. Rice Res. Newal.* 7 (2): 7.

Siwi, S.S., A. Kartohardjono, S. Harnoto and A. Diratmaja. 1987. The green leafhopper genus, Nephotettix, Matsumura. pp. 35-50. In *Proceedings of the workshop of Rice tungro Virus,* Ministry of Agriculture, AARD Maros Research Institute for food crops, Maros Indonesia.

Sogawa, K. 1976. Rice tungro virus and its vectors in tropical Asia. *Rev. Plant Prot. Res.* 9: 21-46.

Suzuki, Y., I.G.N. Astika, I.K.R. Widrawan, I.G.N. Gede I.N. Raga and Soeroto (1992) Rice tungro disease transmitted by the green leafhopper: its epidemiology and forecasting technology. *Japanese Agricultural Research Quarterly* 26: 98-104.

Tantera, D.M. 1985. Present status of rice and legume virus diseases in Indonesia. pp. 20-23. *Trop. Agric. Res. Ser. No. 19,* Tropical Agriculture research Centre, Tsukuba, Japan.

Tarafder, P. and S. Mukhopadhyay. 1979. Potential of rice stubbles in spreading tungro in West Bengal, India. *Int. Rice Res. Newsl.* 4(1): 18.

Thresh, J.M. 1983. Progress curves of plant virus diseases. pp. 1-85. *Adv. Appl. Biol* (Ed. by Coaker, T.H.) 8 Academic Press, New York.

Ting, W.P. and C.A. Ong. 1974. Studies on penyakit merah disease of rice. IV. Additional hosts of the virus and its vector. *Malaysian Agric. J.* 49: 269-274.

Tiongco, E.R., N.G. Fabellar, P.S. Teng and H. Koganezawa. 1992. Tungro viruses in volunteer rice plants. *Int. Rice Res Newsl.* 17(4):20.

Tiongco, E.R., R.C. Cabunagan, Z.M. Flores, H. Hibino and H. Koganezawa (1993) Serological monitoring of rice tungro disease development in the field: its implications in disease management. *Plant Dis.* 77: 877-882.

Tiongco, E.R., R.C Cabunagan, Z.M. Flores and T.W. Mew. 1990. Tungro (RTV) incidence in direct seeded and transplanted rice. *Int. Rice. Res. Newsl.* 15(1): 30.

Van Haltern, P. and S. Asma. 1974. The tungro virus disease of 1972-73 in South Sulawesi. Lembaga Penelitian Pertanian Maros, Sulawesi. Bulletin No. 3: 1-2.

Vidhyasekaran, P. 1985. Forecasting of rice tungro epidemic. *Recent Res. Eco. Environ. Pollution* 3: 119-125.

Vidyasekaran, P. and H.D. Lewin. 1986. Forecasting tungro (RTV) epidemics in Tamil Nadu. *Int. Rice Res. Newsl.* 11(6): 36.

Viswanathan, P.R.K. and M.B. Kalode. 1981. Studies on varietal resistance and host specificity of rice green leafhoppers *Int. Rice Res. Newsl.* 6(3): 7-8.

Wathanakul, L. 1964. A study on the host range of tungro and orange leaf viruses of rice. M.S. Thesis, Univ. of Philippines, Coll Agric. Philippines p. 35.

Widiarta, I.N., Y. Suzuki, H. Sawada and F. Nakasuji. 1990. Population dynamics of the green leafhopper, *Nephotettix virescens* Distant (Homoptera : Cicadellidae) in synchronized and staggered transplanting areas of paddy fields in Indonesia. *Res. Popul. Ecol.* 32: 319-328.

Xie, L.H. and J.Y. Lin. 1982. The occurrence of rice tungro disease (spherical virus) in China. *J. Fujian Agric.* College 3(3): 15-23.

Yadav, B.P. and Mishra. 1990. Spread and infection rate of rice tungro virus in susceptible and tolerant rice cultivars. *Indian Phytopath.* 43: 431-434.